Can Cows Walk Down Stairs?

Can Cows Walk Down Stairs?

Perplexing Questions Answered

EDITED BY
PAUL HEINEY

Illustrations by BILL LEDGER

SUTTON PUBLISHING

First published in the United Kingdom in 2005 by
Sutton Publishing Limited · Phoenix Mill
Thrupp · Stroud · Gloucestershire · GL5 2BU

This revised paperback edition first published in 2006
Reprinted 2006

British Library Cataloguing in Publication Data
A catalogue record for this book is available from the British
Library.

ISBN 0 7509 3748 3

Typeset in 11/17pt GillSans Light
Typesetting and origination by
Sutton Publishing Limited.
Printed and bound in England by
J.H. Haynes & Co. Ltd, Sparkford.

Contents

Introduction

The French philosopher and anthropologist Claude Lévi-Strauss (*not* the inventor of denim jeans), whose complex theories of structuralism are as far removed from my understanding as the small print of an insurance policy, once said, 'The scientific mind does not so much provide the right answers as ask the right questions.' This view, from one of such intellectual authority, comes as a great relief for I've always been able to come up with the killer questions: it's the answers that have eluded me.

I know no more science than I learnt at school, and this was just enough to oblige me to spend the rest of my life frustrated that I did not know more. Somehow, my knowledge of science always stopped two photons short of a good answer to all the fundamental and fascinating questions raised by the business of living. There is a neatness and satisfaction in a convincing answer to a scientific question, but only frustration when a too-slender grasp of the principles involved leaves you unable to produce one. For example, if you ask me why satellites orbit the Earth, I know that it is something to do with Newton's Law of Motion, isn't it? Does angular momentum come into it? Could I define angular momentum? No, of course I couldn't. That's the problem: it's always easier, and much more fun, to come up with the questions than it is to provide full and proper answers.

So I am grateful to Mr Lévi-Strauss for allowing me some credit for only ever wanting to ask why, and expecting someone else to do the hard graft of coming up with the reply. If you are in a similar position, don't worry – Lévi-Strauss puts us at the very heart of scientific thinking.

We occasional seekers after scientific truth are joined by a large body of enquiring people who, in frustration, picked up their phone or connected their computer to the Internet to quiz the brains behind a London-based question-answering service called 'Science Line' – a product of a concerned government which became worried that young people were turning increasingly to the study of media, humanities and sport, and shunning the sciences. This raised the prospect that there might be, a generation hence, no one left in Britain who understood that pi was not part of the food technology syllabus.

The government decided to do something about it. The aim was to provide a telephone and Internet service, available freely to all, which would answer any scientific question anyone, young and old, cared to throw at it. There were a few rules: complexity alone did not rule out any question, so an explanation of why light cannot emerge from Black Holes was fine; but strictly non-scientific questions, such as, 'in the expression, "what's up?", what does the "up" refer to?' was banned. Also ruled out were calls from cheats who were attempting to use the service to do their homework for them.

Behind the scenes at this all-knowing super-brain was a small team of enthusiasts, mostly young scientists from across a wide field of knowledge, who could deal with the recurring questions which were well within the understanding of anyone with a reasonable scientific education, such as 'why is the sky blue?' (Brief answer: blue light gets scattered much more than all the other colours in the light from the sun because of its short wavelength, causing the sky to appear blue.) Of course, very often one question leads inevitably to another, if sometimes out of a natural desire to prove yourself smarter at asking questions than the other person is at answering them. Such a clever-clogs might now ask, 'if the sky's blue because of scattering, why is a sunset red?' (Brief answer: because the light from the sun, as seen at sunset, is coming to you through a thicker atmosphere, which absorbs the blue light. But the blue light of the sky doesn't come directly from the sun – it's scattered light.) And from a fertile mind a hundred further questions can probably emerge, but we'll draw the line there.

Some questions asked of Science Line, however, were harder for them to deal with than the blue sky theory. 'What is the exact difference between Henkin's proof for the Completeness Theorem for First Order Logic and Godel's proof?' Eh? I'm sorry; I would need someone to explain the question before I could begin to understand any answer. But Science Line was not fazed by this, nor by questions like 'can you describe a method for determining electron dn configurations?' – which sounds suspiciously like homework to me. Instead of scratching their heads, the experts forged contacts with the wider academic community and turned to them for definitive answers. As a result, all the answers carried authority, readability and sometimes no small measure of humour.

Then, just as Science Line was becoming part of people's lives, the government pulled the funding plug and it passed peacefully away. Its website carried the sad statement, 'Due to lack of funding Science Line will close on the 26th September 2003. We are sorry but we can no longer take any questions.'

By great good fortune, before the website closed and the dedicated team of question answerers moved on to other things, they had already explored the possibility of a book based on their vast, wide-ranging database, which by now contained over 16,000 questions and answers. This is where I came in, although at this early stage I was entirely unprepared for the breadth, depth and often sheer entertainment value of the material Science Line had collected. I have always thought that the idea of the undiscovered treasure chest which opened to reveal jewels and sparkling gems was the stuff of children's books. But now on my desk were two slim computer disks, which opened to reveal equally dazzling contents. Together those two slim disks added up to a mountain of knowledge which we all agreed should not go to waste. The questions and answers would no longer be on the Internet, but why shouldn't the best of them be brought together in a book?

I hadn't read far into the megabytes of knowledge before I realized that here were the answers to questions that had dogged me all my life. I already knew why the sky was blue, honestly, but I had no idea why flies circle round light bulbs, or why jelly made with fresh pineapple will never set – but I do now. I understand reflections in mirrors now – did you know they're not 'flipped' at all? And if you have ever lain awake at night wondering whether or not penguins have kneecaps, the answer is here. You will also learn why cows can walk up stairs but not down again – that's to do with kneecaps too.

There can be no greater pleasure than sifting through these questions, not only for the satisfaction of discovering the answers, but for the sheer enjoyment of the lateral thinking and mischievous minds which asked, 'how easy is it, scientifically, to fall off a log?' or 'do bacteria have sex?'.

The task of choosing the questions to include in this book was an easy one – I picked the ones that not only fascinated me most, but also those which provided surprising or unusual answers. The selection was made purely on entertainment value: the sort which leads not to a giggle or a belly laugh, but to the warm glow that flows from a nagging scientific question answered in an understandable way. No doubt someone else would make an entirely different selection.

The questions, remember, belong to the people who asked them, and for the education they provide we must thank them. And to those who patiently answered them I can only express the most enormous respect for what they clearly believed was a vital public service. I herewith salute and give full credit to Siân Aggett (Biology), Alison Begley (Astronomy and Physics), Duncan Kopp (author of *Night Patrol*), Khadija Ibrahim (Genetics), Kat Nilsson (Biology), Jamie McNish (Chemistry), Alice Taylor-Gee (Chemistry) and Caitlin Watson – as well as the numerous distinguished experts whose knowledge they drew upon when their own was stretched to its limits.

On occasions I have reworded their answers for clarity, and sometimes added to them if I thought further explanation was needed. But this book is really the work of those who asked the questions and the dedicated few who answered them.

I hope that after reading it we might be able to convince Mr Lévi-Strauss, were he still alive, that mastery of both the question *and* the answer leads to the supreme scientific mind.

Paul Heiney
2005

1

Where it All Began: Secrets of the Universe

Atoms to Big Bangs

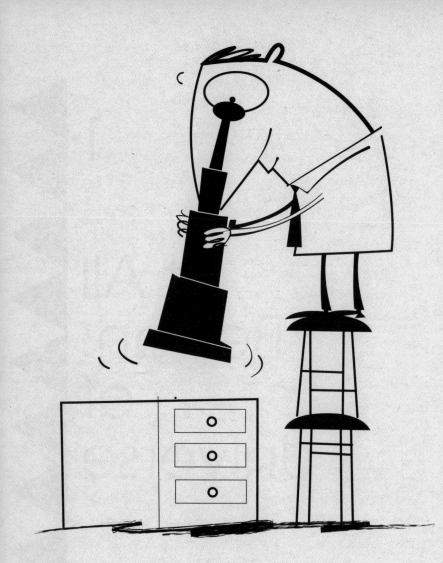

What does
an **atom**
look like?

That is very difficult to say because they are so small and we cannot see them even when we use the best microscopes we have. But scientists are now using a new kind of microscope to make images of atoms: these machines still can't *see* the atoms, but they can *feel* them in a similar way you feel a prickly sensation when you hold your palm very close to the TV screen but don't actually touch it. This is complex nanotechnology but, clever though it is, it still doesn't let you see an atom. If you could, you'd find there is a tiny core at the centre called the nucleus, consisting of particles called protons and neutrons. Protons and neutrons have about the same mass. Protons have a positive charge. Neutrons have no charge.

Hydrogen was the very first atom to be created at the moment of the Big Bang – the start of the universe – and it was created by the coming together of a quark (yet another sub-atomic particle) and an electron. The Big Bang theory of the creation of the universe occupies many volumes and some of the finest minds, but in a nutshell it says that the universe began with all its matter concentrated at very high density and temperature 15 billion years ago. An explosion caused it to expand, which it is still doing to this day.

If **atoms** are mainly **space,** why doesn't your **hand** fall **through** a table?

Everything around us is made of atoms — even the air we breathe. The difference between the air we breathe and, say, a table is that the atoms are much more tightly packed in a table. So while you can pass your hand through the air — where you are essentially pushing atoms out of the way — you can't put your hand through a table because the atoms can't move out of the way. It's like trying to walk through a tennis court filled with 100,000 tennis balls as opposed to one with 100 tennis balls: you just can't do it. But it's not just the space that's a problem. There are also very strong forces holding atoms together. So although atoms *are* mainly space, the strong forces holding them together and the tightness with which they're packed mean you can't put your hand through a table. It's not the space within the atoms that prevents you, it's the forces that hold them together.

I've **heard** that
if you remove all
the **space** from **around**
the **atoms then** all the atoms
in the universe would **fit into**
a **matchbox.** Is this true?

This is one of the tales you hear now and again. I could simply tell you it's not true, but it would be better if we worked it out. It won't be an accurate calculation, but it will give us an indication as to whether there's any truth in this 'universe in a matchbox' theory.

This is how it goes:

1. We need to know how much space the bits in an atom will take up. For simplicity, let's assume that there are only hydrogen atoms, so we need to work out the volume of a proton and the volume of an electron. Assuming that both the electron and the proton are spheres, their volume will be given by the formula $V = 4/3$ pi r^3

    Volume of electron:

    radius is approx $= 2.82 \times 10^{-15}$ m

    so volume works out as approx $= 1 \times 10^{-43}$ m^3

    Volume of proton:

    radius is approx $= 1 \times 10^{-15}$ m

    so volume works out as approx $= 2 \times 10^{-44}$

    This means that the total volume taken up by the proton and electron is approximately 1×10^{-43} m^3

2. We now need to know the volume of a matchbox, which is approximately 3×10^{-5} m³.

3. The next step is to work out how many atoms can fit in this matchbox. This can be calculated by dividing the volume of the matchbox (3×10^{-5}) by the volume of the atom (1×10^{-43}), giving the answer 3×10^{38} atoms.

4. The final step is to see how this number compares with the number of atoms in the universe. There are a couple of ways of estimating this. There are about 100,000,000, 000,000,000,000 stars in the universe – and that's a guess. How many atoms in a star? Impossible to say, although we can have another guess. Let's say the sun is a typical star and it's made up totally of hydrogen. The mass of the sun is 2,000,000,000,000,000,000,000,000,000,000kg. The mass of a hydrogen atom is 0.00000000000000000000000000 017kg. Divide one by the other and the number of atoms in the sun is 1,200,000,000,000,000,000,000,000,000,000, 000,000,000,000,000,000,000,000,000.

    Now multiply that by the number of stars in the universe and you have: 1 with 77 zeros after it as the number of atoms in the universe.

Looking at it another way, the mass of the observable universe in kilograms is 1 with 52 zeros after it. That's about 90 per cent of the total mass of the universe, we think, so the total mass of the universe in kilograms is 10 with 52 zeros after it. Divide that by the mass of a hydrogen atom (the vast majority of the universe is hydrogen) and the number of atoms is 6 with 79 zeros after it.

These two figures are similar enough to suggest that they're about right, so we will take as an educated estimate of the number of atoms in the universe.

5. Comparing the answer to point 3 with the answer to point 4, it is clear that even if all the atoms in the universe were as small as a hydrogen atom they wouldn't fit into the matchbox. Since there are many atoms that are much larger than hydrogen the actual number that will fit in the matchbox will be smaller than we worked out in 3.

So how much volume would all the atoms in the universe take up if they were all only made of an electron and a proton and no space? This can be worked out by multiplying the answer to 4 by the answer to 1:

$$1 \times 10^{79} \times 1 \times 10^{-43} = 1 \times 10^{36} \text{ m}^3$$

A rather big matchbox!

What
is **time?**

Do you want a psychologist's answer, or a physicist's? I suspect the latter, in which case you have to prepare yourself to confront the theories of Albert Einstein, one of the most important scientific thinkers of the first half of the twentieth century who established the theory of relativity.

According to Einstein, time and space are closely tied together and he went on to show that asking *when* an event takes place is much the same thing as asking *where* it takes place. He said that we could not separate the world into space and time but that space and time are actually parts of something called space-time. Space-time has four dimensions: three to give a position in space and one to give a position in time. When you walk, you move through space-time; when you stand still (because time passes) you are also moving through space-time. Our experience of time is a result of moving forward in this time dimension of space-time. So time is really another dimension, the difference being that in the other three dimensions we have a choice about which direction we want to travel in. But with time there's only one direction and that is forward — except in the case of Dr Who.

Was there **a time** when time 'began'? And **what** was there before the **beginning of time?**

If you believe the Big Bang theory of the creation of the universe (see p. 13), then the point of the explosion is the moment at which time began. If you want to put a figure on it, you could say it happened about 15 billion years ago when, according to the theory, all matter, space, energy and time were created from a single point called a singularity. At this point $T = $ zero. To answer the tricky question of what was there before the Big Bang, cosmologists have found a sneaky way of rendering the question meaningless – that way they don't have to give you an answer. They argue that T cannot be a negative number because there's no such thing as negative time, so there's no point asking what happened before $T = $ zero. A useful analogy is to imagine yourself at the North Pole asking the question 'which way's north?'– the question doesn't make sense.

What is the **main cause** of gravity? **Why** is there an attraction **between two** masses?

Sir Isaac Newton in the seventeenth century first formulated a law of gravity: it said that two masses will attract each other with a force which depends on their distance apart and their masses. It was one of those laws which came about through observation and experimentation, and only then was it set out mathematically. As to what actually *caused* the forces of attraction, not much thought seems to have been given to that at the time.

It had to wait until Einstein in the twentieth century turned his attention to gravity. He equated gravity with the acceleration of a body and went on to show that light bends in a gravitational field. Since light has no mass, Newton's theory could not explain this bending. Einstein's great contribution was to show that space-time actually curves owing to mass. Imagine a heavy ball sitting on a large, stretched rubber sheet – space would curve near the mass but remain reasonably flat further away. Only if light passes close to the large mass will its path be appreciably deviated. Experiments have now been performed which show that light really does bend near a mass owing to the curvature of space-time.

But that's not really the answer to what gravity actually is, and no one has yet come up with a theory which enables us to describe it.

ISSAC NEWTON

I know that the **length** of the days and years is **due to** the **spinning of** the planets on their own axis and **their orbit** round the **sun,** **but what started them** **spinning** in the first place? What got them going?

To answer the question you have to go back to the formation of the solar system which was created out of a massive ball of gas and dust which slowly started to pull together under the effects of gravity. As this dust came together, particles colliding as it did so, the centre of the ball got hotter and hotter until it became hot enough to form what we call the sun. As the temperature rose, the sun reached a point at which it became 'switched on', like a fire suddenly igniting. This ignition caused gas and dust to be flung away from the sun to form the basic building materials of the planets.

Now for the spinning. There is a law of motion called the 'conservation of angular momentum' which says that as something gets smaller, it spins faster and faster. This is why, for example, a skater speeds up if she brings her arms into her body making herself smaller. It's the same with a ball of dust and gas: any slight rotation which it already had would have become bigger and bigger as it decreased in size. Now, as things spin, centrifugal forces push the middle out and pull the top in. This happened with the ball of dust so that eventually it wasn't a ball any longer, but a disc surrounding the sun. The planets then formed from this disc, which is why they all orbit in roughly the same plane around the sun.

That original ball of gas wouldn't have needed much spin to produce the rotation we see in the solar system, although what produced that original spin is not known. But things in the universe generally like to spin if they have any choice. Virtually everything from galaxies to planets spins.

Does light **stop** existing, or would it **travel on** for ever **into** eternity if **nothing** got in **its way?**

The answer lies in the words 'if nothing gets in its way'. In theory, light will go on for ever if it doesn't bump into anything, but that requires it to travel through a perfect vacuum, which in practice can never occur. Light is energy, and if nothing occurs to make it lose that energy, then it will exist for ever.

Imagine a photon, which is a parcel of light emitted from the sun. If it manages to miss all the planets and asteroids and comets (in other words all the large objects in the solar system), it may well hit a piece of dust from a comet, or a mere atom of hydrogen just floating around in space, and lose its energy that way. But some photons survive the journey and travel in a straight line until they meet, say, your eye. That is then the end of that parcel of light, for the energy the light was carrying is converted into an electrical signal that goes to your brain, allowing you to see the light.

Alternatively it might collide with an atom floating in space, or an atom in the atmosphere of a planet, or an atom in an object, like a rock. Some of that energy gets reflected – and that's how we see things.

This **Big Bang** sounds awesome.
Was it actually
a bang,
as in an explosion?
And **would you**
have heard it
if you'd been there
at the **time?**

This, of course, is a hypothetical question with no good answer to it. But how about this for a theory? Sound, which is transmitted as vibrations, needs something to travel through. At the time of the Big Bang, it's true that the universe was infinitely dense, but there weren't any individual particles, so I imagine that sound wouldn't have travelled. But if you've got a better theory, that might be right too.

Could you ever travel **fast** enough to **overtake** the Big Bang? **I mean,** if you travelled at **twice** the speed **of light** would you **overtake it,** and then **be able** to watch the start of **the universe?**

Sorry, but even if you did get to twice the speed of light, you have to remember that the Big Bang created not only the matter in the universe, but also the space within it. How does that stop us looking back at the creation of the universe? Just after the Big Bang the universe was still tiny, just a few metres across. If we tried to travel out of it, we wouldn't have anywhere to go because the space hadn't been created.

Is it **possible** that there was **more than one Big Bang** and that there are, in fact, other **universes** moving **towards** one another?

There's no way to tell. First of all, the concept of an expanding universe is a tricky one and it's often misunderstood. The universe is not expanding 'into' space: it's not as though there's something out there that is slowly being filled by the expanding universe — it is space itself which is expanding.

In other words, the distance between two objects in the universe is getting greater, but the objects are not moving. That's why you can't have two Big Bangs next to each other.

Are you **saying** that there's **nothing** outside the **universe?** **Surely** it's **got** to be **contained** in *something*.

Some of these questions are for scientists and others for philosophers. This question is largely for the latter. Technically, 'the universe' means *everything*, and so there can't be anything else beyond it otherwise that would be part of the universe too. I think where the confusion arises is because we use the word universe to describe everything we can see, when we should really be describing it more accurately as a 'visible universe'. There is, of course, much beyond it which we can't see because there hasn't been time for the light from far-away bodies to reach us. The universe is around 15 billion years old, so we can see anything within a distance of 15 billion light years from us, because the light from these bodies would have travelled to us. The universe has not been around long enough for the light from anything beyond this distance to get to us.

As to the universe beyond what we can see, it's guesswork to some extent. We can tell what it *might* be like because it will have a gravitational effect on us, even if we can't see it. Einstein's equations of general relativity, which describe how gravity affects space itself, are still the best way we have of describing our universe on its largest scales. These imply that the space is either infinite, or is closed back on itself. If it's infinite, then it can't be contained in anything because it goes on for ever; if it's closed it doesn't really have a beginning or end either. That's kind of hard to think of in three dimensions, but imagine you are two-dimensional and wandering around on the surface of a sphere. You can move back and forward, and left and right, but you don't have any concept of up or down. As far as you're concerned, there is nothing else but the surface of your sphere. Then, you would be forever wandering around on this sphere but never come to an end of it. So, for us, our universe is all there is.

This **space** that the **universe** is expanding into, what **is it?** Is it pure **emptiness,** nothingness? If I **took** an open box out into **space,** closed the **lid** and **brought** it back to **Earth** what **would be** in it?

Space is not a perfect vacuum. Even if we could somehow get rid of all the interstellar dust, etc., at the quantum level space is not empty – it consists of shifting quantum fields, which are apparently due to the universe's gravitational field. So your box wouldn't be filled with nothing. Space is not really 'pure distance': it is the name we give to the (almost vacuum) surroundings containing all the galaxies, and describes the gravitational field of the universe. This is something we still don't fully understand, so a 'playground for the celestial bodies' seems as good a term as any!

What are the so-called black holes in the universe?

John Michell, an English astronomer, first suggested in 1783 that a mass could create a gravitational field so strong that light could not escape from it. A few years later the French mathematician and philosopher Pierre Laplace arrived at the same conclusion. Then, when Einstein proposed his theory of general relativity in 1915, a black hole as a real object became a possibility. John Wheeler coined the term black hole in 1967.

There is no absolute proof that any black holes exist, but there is evidence for them. The first black hole to be 'discovered' was Cygnus X-1 in 1971. Although no one can say for certain that this is a black hole, few people now doubt that it is.

But why can't **light escape** from a **gravitational** field? **Light** doesn't 'weigh' **anything,** so **what** holds it back?

Explaining black holes is very difficult if you simply stick to Newton's ideas of gravity. They work fine when we talk about everyday activities like playing pool or throwing balls – even launching rockets works with Newtonian gravity. But when it comes to complicated things like black holes, you have to start looking at what gravity does to space. This is what Einstein was doing in the early twentieth century. His theories of gravity say that gravity affects a combination of space and time called space-time. Einstein said gravity bends space-time so that light doesn't travel in straight lines. The quickest way from A to B is always a straight line, unless it isn't!

This might help you understand: you'd think that planes flying from London to Vancouver on the west coast of Canada would simply fly straight across the Atlantic, but they don't. They fly north towards Scotland and then head over Greenland because this is actually the straightest and shortest route, although it doesn't look like it. Our normal view of the world is a flat one – all the maps we use are flat – so it looks as though the shortest route is straight across the ocean. But if you look at a globe – a true representation of the world – you'll easily see that the shortest route is what's known as a Great Circle over Greenland.

It's the same with space-time. Our view of space is that it is flat, and that view is perfectly acceptable as long as all we want to do is go to the moon. But as soon as we start talking about areas of space where the gravity is very strong – black holes, for example – we have to start taking into account the effects of gravity on space-time. Imagine a trampoline with a grid of straight lines drawn on it. If you put a heavy sack of potatoes in the middle of it, the trampoline will sag into the middle and the straight lines won't be straight anymore. If you then roll a marble from one end of the trampoline to the other, it will no longer go in a straight line but will follow the bent lines on the trampoline. And that's what happens with space-time and light. Gravity bends space-time, and light follows the straight lines that have been bent out of shape through it. A black hole bends space-time so much that the straight lines actually bend right round on themselves and the light ends up going round and round in circles. That's how black holes work.

What would happen if I fell into a black hole?

The first thing to understand is that you're never going to get out again. As you approach, you won't feel much at all. Like an astronaut orbiting the Earth, you'll be in free fall and every part of your body will be under the same gravitational forces

– you'll feel weightless. But once you start to get closer to the immense gravitational field of a black hole, about half a million miles from the centre, you come across what is known as the black hole's tidal force. If you happened to be heading towards the hole feet first, your feet would feel more gravitational pull than your head and you would suffer the sensation of being stretched. This would get worse to the point where your body 'twanged', and that would be the end of you.

Most likely, this would happen before you crossed something called the black hole's 'horizon'. At this point the speed at which you would have to be travelling to make your escape becomes equal to the speed of light. All gravitational fields have an escape velocity; on Earth it's the speed at which a rocket has to travel to get into space. Once you cross the horizon of a black hole you need to travel faster than the speed of light in order to escape, which is impossible. Once you're over the horizon, you're trapped, if you haven't already been stretched beyond endurance.

What would I see as I **fell** in?

Things may look a little distorted since the light from distant objects would be bent by the huge gravitational field, but even when you've crossed the inescapable horizon, light from outside can still be seen from the inside. Of course, no one can see you because light from you can't escape from the black hole – to do that, light would have to travel faster than light, which is clearly impossible.

The next stage of your journey takes you towards the 'singularity', which is the centre of the black hole. You are now in a strange world where distance has become time and there's no possibility of avoiding the singularity because it is no longer a place to which you are heading, but a time within your future. There's no escaping it, like there's no way of avoiding the day after tomorrow – it's going to arrive whether you like it or not.

Why do black holes happen?

They are created when massive stars collapse in on themselves because they have run out of fuel to keep them burning. Stars are made of gas and they work by converting one gas into another – usually hydrogen into helium. Eventually most of the hydrogen is converted to helium, then the helium is converted to carbon and carbon changes to oxygen. All of these reactions release energy in the form of heat and light, which makes a star hot and bright. This light and heat maintain the shape of the star, preventing gravity from pulling all the gas to the middle.

But there comes a stage when there just isn't any more fuel to burn, and gravity takes over. If the star is big enough (it must be more than three times the size of the sun) it will collapse inwards. Then the density of the material in the centre becomes so high that the gravity of the object is great enough to stop light escaping. It has become a black hole.

I want to **try** to **get the hang** of the distances involved in this **universe** of ours. **How long** would it take me **to travel** to the limits of the **galaxy** using current **technology?**

You'd never get there. Not because you'd die before you got to the end of the journey, but because the journey itself has no ending. The popular theory today is that the universe is expanding and will continue to expand for all time, and galaxies at the far side of the universe appear to be receding from us at very close to the speed of light because of this expansion. So with today's technology (the space shuttle travels around 28,000km/hr) you could never catch up with the edge of the expanding universe. It's a race you'd never win.

Having said that, there is not really going to be an edge of the universe to get to. If the universe is curved, as some theories suggest, then it will fold back on itself forming a shape without any edge, like the surface of the Earth does. If you travel outwards in one direction on the Earth you will come back to the beginning. The same *may* be true of space – that if you travel for long enough in one direction you will come back to where you started. If it turns out that the universe doesn't fold back on itself, you're still not going to get to the edge because then the universe would be infinite.

For fun, let's forget about the expansion and shape of the universe and board a space shuttle and head at 140,000 km/hr for the furthest object we can see, which is about 10 billion light years away – or 95,000,000,000,000,000,000,000 km. A quick stab at the calculator tells you your journey time would be 75,000 billion years. While contemplating this, remember that the universe isn't much more than 15 billion years old.

How do you know how far away some of these stars and galaxies are? How do you measure the distances?

First, you have to understand an effect called parallax. If you hold up a finger in front of your nose, about 20cm away, then close and open your eyes alternately, the finger looks as though it's jumping from side to side. This is because each eye gives you a different view of it, and your eyes are a few centimetres apart.

If you know two vital measurements: the distance between your eyes, and the angle that your finger appears to jump, then a bit of trigonometry will tell you how far away your finger is from your eyes.

The problem with this method is that it's fine for close-up fingers, but with objects further away, the movement of the distant object is very small. If you were to try this with a lamp-post at the far end of the road, you'd find you wouldn't be able to detect any movement of the post at all – it would be too small to observe. So, to increase the parallax the 'eyes' have to move further apart. Astronomers create this effect by making one observation at a point in the earth's orbit, and then waiting for the earth to go halfway round its orbit (which takes roughly six months) before taking a second observation. By having moved the 'eyes' apart by twice the distance of the Earth from the sun, the base line is large enough for the distance to be measured to stars as far away as several hundred light years.

Is there **much junk** in **space?**

Actually it's getting quite busy up there because there are increasing amounts of man-made objects, and as these collide with one another, that creates yet more debris. At a rough estimate, there are 7,000 large objects in space, about 300–500 miles high. Some 2,000 of these are payloads, but only about 5 per cent of them are working. Then there are 40,000 bits and pieces which are the results of impacts or the remains of exploded rockets. You can add to that about 3 million particles which might be flakes of paint, insulation or dust, some of which are moving at speeds of up to 18,000mph – fast enough to craze the windows of the Mir space station.

If **everything's** weightless in space, how do **spacemen weigh** themselves?

If I told you that they did it by shaking themselves around you'd say I was kidding, but it's true. You have to understand this: weight is a force by which a body, say a spaceman, is attracted to the Earth. If you took him into outer space, where there might be no gravitational forces, then he would truly weigh nothing. But he/she will still have mass because mass is a measure of the amount of material an object contains. Of course, weight and mass are linked; weight is the product of mass and gravitational attraction so the greater the force of gravity, the greater the weight although the mass remains unchanged.

To measure mass in space you have to use a device which works independently of gravity: it's called an inertia balance. Remember, your inertia is also a measure of your mass – or the more 'massive' you are, the more difficult it is to get you moving. So astronauts strap themselves to a shaking device, the inertia balance, which jiggles them backwards and forwards, working out how much effort is required to get them moving. From this the mass of the astronaut is calculated, and that would give you his equivalent weight on Earth.

If everything's **floating around** inside a **spacecraft,** and I mean *everything,* then precisely **how do** astronauts use the **toilet?**

There are things about space toilets that you would find very familiar. It looks like an ordinary toilet, suits both men and women, has a light for reading and boasts a window through which the seated astronaut has a fine view of Earth. What you might not be so familiar with are the straps, foot restraints and seat belt. Now, to business.

Early space suits used nappies and disposable bags, but nowadays it feels pretty much like a normal toilet operation. The main difference is that there is no water flush. Instead, solids are moved by a blast of air into a compartment (out of sight) where they are dehydrated, disinfected, compacted and stored for disposal after landing. The liquids are released into space and vaporize. The air in the toilet is cleaned, filtered and conditioned before being pumped back into the cabin.

There's an even newer system where plastic bags are placed in the bottom of the lavatory and which catch both solids and liquids and seal them, stacking them one on top of the other as the toilet is used. This method has overcome the problem by which the extractor fan corroded as a result of contact with urine.

Suppose it **happened** to be my birthday while I was **out** in space, what would **happen** if I tried to **light** a candle?

You're right to be interested in candle flames: the great nineteenth-century scientist Michael Faraday said, 'There is no more open door by which you can enter into the study of natural philosophy [science] than by considering the phenomena of a candle.'

I'm assuming you're going to try this in the spacecraft and not in space itself. On Earth, the beautiful shape of the candle flame is formed by the burning of the wax in the presence of oxygen to produce, among other things, carbon dioxide and water. These rise from the flame, and oxygen from the air is drawn in to replace them. That's what gives the flame its shape.

In a spacecraft the flame exists in microgravity, the hot gases don't rise and fresh oxygen isn't brought in from below. The result would be a spherical blue flame, which wouldn't last long because without oxygen the wax can't burn.

Suppose I got to **Mars,** would there still be **Christmas?**

Actually, you wouldn't feel too uncomfortable about the passage of time if you ever got to Mars, because at 25 hours to a day the Martian day is pretty similar to ours. But the year would be longer because Mars takes 687 days to complete an orbit of the sun. Looked at that way, you'd only get one Christmas roughly every two Earth years. However, if you decided that Christmas comes around every 365 days, then on Mars Christmas would come twice a year. Enjoy!

Saturn has rings which you can see from Earth. Why doesn't the Earth have rings? What's so special about Saturn?

Saturn's not the only planet to have rings: Jupiter, Uranus and Neptune have them too, but they not visible from Earth, unlike Saturn's. It is only since the expeditions of the spacecrafts Voyager 1 and 2 that we've known about them. What's interesting about the rings is that all the gas giants, as the outer planets are called, have them and astronomers now believe that all the rings round the outer planets may have been formed in the same way. There are two theories: the first says that the rings are made up of rock and dust that was formed by asteroids colliding near the planet. The gravity of Saturn and its moons has since trapped the dust and rocks into the rings we see today. The second theory suggests that when the planets were forming from the dust and gas cloud, not all the dust and gas was collected by the planet. In other words, the rings are simply the leftovers from when the planets formed. Now, if astronomers could find out how old the rocks in the rings were, they might be able to tell which theory was right. A lot of people think the first theory is correct because Jupiter, Uranus and Neptune have such faint rings. They say Saturn's rings are only bright because they were formed 'recently' – which in astronomical terms means millions of years ago – from asteroids colliding. The other planets' rings aren't as bright because they formed a long time ago and most of the bits in the rings have been sucked into the planet.

Why doesn't the Earth have rings? You need a source of material to form rings around a planet, and that material must not be too far away – no more than three times the radius of the planet, which is much closer than the moon. In the case of Jupiter, it appears that the dusty rings might be made up of material blasted off Jupiter's much closer moons by meteorite impact.

The other factor to take into account is the power of the solar wind. This is a constant flow of energy outwards from the sun, which, because of our closeness to the sun, has a far greater impact on Earth than on more distant planets. This wind would easily sweep away any small particles that tried to orbit the Earth.

Even if the Earth did have a source body to provide rings around it, they'd be pretty dark and dusty because any bright icy material (of the sort which makes up Saturn's rings) would be evaporated by the heat from the sun. Another possible reason why they wouldn't last very long is that the solar and lunar tides are pretty strong, and would certainly disrupt the ring system eventually. We might have rings for a short period of time if we captured a small asteroid and it ended up in orbit at the right distance, but they probably wouldn't last for long.

Is it true that a **tenth planet** has been discovered, **even further** away than the **gas giants?**

There is lots of stuff floating around the solar system in orbit round the sun. The problem is, where do you draw the line between something big enough to call a planet, and something better classified as debris? There are some who argue that Pluto isn't really big enough to be described as a planet, although it is largely accepted as one.

However, if Pluto makes it then so should the planet Sedna, named after the Inuit goddess of the seas. Sedna is 90 per cent of the size of Pluto, with an estimated surface temperature of minus 200°C, and occupies a position roughly 10 million miles from the sun. It was discovered in November 2003 by the Palomar Mountain telescope in southern California, and since confirmed by other observers. It orbits the sun once every 10,500 years and it is thought it may have a moon. The team that discovered the tenth planet expects to find others as observations continue.

If **astronauts** ever did get
to the planets, how would they
navigate? **GPS** wouldn't work,
and I imagine navigation by **stars**
wouldn't work either.

You're right about GPS, the Global Positioning System, which works through man-made satellites in orbit round the Earth. If we ever got to Mars, we'd find that it has a North and a South Pole, just like ours, but the magnetic field is 800 times weaker. So, with sensitive enough compasses, you might be able to find your way round Mars. Strangely, if you wanted to astro-navigate using the sun, planets and stars, as mariners have been doing for a couple of centuries, then that would work fine. The night sky seen from Mars would look similar to that viewed from Earth, and by taking measurements of stars, and knowing the time, you would be able to fix your position to within 100 metres or so on the surface of Mars.

What is
the **moon** made of?

The moon is a quarter of a million miles away from us and it condensed out of a swirling cloud of rocks and gas about 4.5 billion years ago when the solar system was formed. Of the nine planets orbiting the sun, many have their own moons spinning round them, just like us. Some have more than one moon: Saturn had seventeen at the last count. But the Earth's moon is the biggest in the solar system.

Scientists used to think that it was a huge piece of rock that had been torn out of the Earth leaving a large hole, now filled by the Pacific Ocean. That theory has gone out of fashion and it is now thought that the moon probably condensed out of the swirling gases as a separate mini-planet. It was then captured by the gravitational pull of Earth and became our moon.

Before we landed robots and eventually humans on the moon in the 1960s we were unsure of its composition. Lunar samples which were brought back show that it is a volcanic rock, basalt, similar to many of the volcanic rocks on Earth. Basalts form on Earth when volcanoes erupt, throwing out molten rocks into the air or the sea. These hot rocks (initially at thousands of degrees centigrade) cool very quickly to form dark rocks with small crystals – basalt is a typical example.

Basalts are made of four main elements. They are mainly silicon, but they also contain iron, aluminium and magnesium. Silicon is the most abundant element on the Earth and occurs in many rocks. Sand on a beach is largely silicon. Iron, aluminium and magnesium are all familiar metals.

Inside the moon there is a crust, a mantle and a core, just like the Earth. But the moon has cooled down a lot more than the Earth and its mantle is no longer molten so there are no active volcanoes on the moon any more. There are, however, occasional earthquakes, or moonquakes as they are better known.

Do we *need* the **moon?** If the moon **disappeared**, would **we** survive?

Actually, the moon is moving away from us, but hardly fast enough for us to be worried about it. The distance between the Earth and the moon is increasing by 3.82 centimetres a year. I doubt you notice it.

But if the moon were to suddenly disappear, that would be a different matter. For a start, the tides across the globe, caused by the gravitational pull of the moon, would cease to happen. That would have far-reaching implications for sea-borne trade. But that apart, there would be no reason life couldn't continue pretty much as before.

However, there is a suspicion that the tilting of the Earth's axis may be controlled by the presence of the moon, and if that influence were to be removed then the possibility arises of dramatic changes to the lengths of day and night, and to the cycle of the seasons.

What keeps the **moon** up there? I **watch it** every night and **there it is**, up in the sky. **What** stops it falling to **Earth**, **attracted** by our gravity?

The moon *is* actually falling down, but it is moving to the left as well, if you look at it from the northern hemisphere. For each distance that it falls, it also moves out to the 'left' and misses the Earth. So, it keeps falling while at the same time moving to the left, and falling and moving to the left, until it gets back to where it started — a full orbit. So the moon *is* actually in free fall, it just keeps missing us.

If **you** were able to **write**
in the dust
on the **moon's** surface,
how big would the letters
have to be
in order **to be visible**
from **Earth?**

You're talking pretty big letters. If you draw a line from one side of the moon to an observer on Earth, and then back to the other side of the moon, you will form an angle of about 8 degrees. (Incidentally, the sun also has an angular diameter of about half a degree which is why we get such perfect eclipses.)

All telescopes have something called an angular resolution which is the smallest angle they can see. If a telescope has an angular resolution of, say, 1 degree then it would not be able to see the difference between an object which is 1 degree across and one that is 0.5 a degree across. The hugely powerful Hubble space telescope has a resolution of about 0.1 arc seconds, which is equivalent to seeing things which are

To work out the size of an object that would be visible to an observer, you have to start with the distance apart. The moon is 384,403 kilometres away. If Hubble is at its closest point to the moon then the moon is about 383,800 kilometres away from it. By using trigonometry we can see that at this distance the smallest object that Hubble can resolve (remembering that it can see things a mere 1/36000 of a degree across) is about 200 metres.

A human eye, of course, is far less powerful and can only see things greater than 1/60th of a degree across. Therefore, the smallest object the eye could see on the moon would have to be 110 kilometres across.

What if
the sun
disappeared?

For the first eight minutes after switch-off, we would be blissfully ignorant of the fact that the sun had gone out. Soon after that, things would get serious.

Eight minutes is how long it takes for light and gravitational particles from the sun to reach us – light travels at 300 million metres per second and the sun is 150 thousand million metres away from us. Divide the distance by the speed to get the time – which is 500 seconds, or about 8.3 minutes.

After that, our orbit would begin to change as we would no longer have the sun to revolve around; instead of moving in a circular path the Earth would probably move in a straight line, but it is difficult to be certain. So, the Earth would be plunged into darkness and start to veer off into the universe bound for who knows where?

It is doubtful whether an instant freeze would take place since the Earth has absorbed a lot of heat from the sun and has its own hot molten core of iron at its heart, as well as an atmosphere which acts as a blanket, so the cooling of the planet might take some time. It might start off rather like the coolness experienced after sundown, but thereafter the temperature trend would be steeply downwards.

What would be most catastrophic would be the loss of light which plants need to use for photosynthesis: food crops would cease to grow and plants that feed animals would quickly die. The animals would starve. There are plenty of life forms which are capable of surviving without light – for example, chemo-autotrophic bacteria and certain ocean-dwelling creatures (tubeworms which live near thermal vents) – and these will outlive humans, although it's difficult to say for how long.

It is also difficult to predict how the oceans would behave as the impact of the moon's gravity on tides might be greater once the sun is no longer the strongest gravitational force. There again the moon might fly off on its own, leaving our planet tide-less, although I suspect that would be the least of our worries.

If the **sun** did suddenly go out, **would** we **see** it or **feel** it first?

Even in the most energetic of explosions – assuming that's how the sun would expire – any particles which fly out will always travel more slowly than the light. So there would definitely be no effect from any of the particles before the darkness arrived.

As far as feeling the absence of the sun, the radiation which heats the atmosphere arrives in the form of infrared which travels at light speed (since it is just low energy light). So the infrared radiation arrives, does its job and then we feel the effect some time after. Because of this delay, it is thought that the sun would have to have 'been out' for about a week before the earth started to freeze. So you would most definitely experience the darkness quite some time before the difference could be felt.

Does the **sun** have a lifetime? If it eventually **reaches** the end of its **life**, **what** will happen?

Yes, the sun has a lifespan, but it is thought that it will continue with business as usual for another 5,000 million years until it is roughly twice its present age. For all this time it will generate energy by nuclear fusion of hydrogen – combining several hydrogen atoms to create helium and releasing energy. But gradually, this helium will become the dominant element in the core and eventually all the hydrogen there will be used up. The sun will then pass from middle age to old age.

Hydrogen burning will then move out into a shell around the core, gradually spreading out through the sun and using up fuel as it goes. This creates instability inside the star, making it swell to a vast, cool red giant. By now it will be large enough to engulf the Earth's orbit. At this stage a reversal takes place as the temperature and pressure inside the core build, eventually reaching a point where helium fusion can begin, generating energy in the core again. This return to normal service will be short-lived, for after a few million years the helium will have been consumed.

Eventually the helium burning will follow the pattern of the hydrogen burning, moving out into the shell, and the sun's internal pressure will overcome gravity again, and it will swell back to a red giant.

But this time the sun is unable to generate enough energy to begin burning the heavier elements at its core, and it really will be the end. The expansion will continue, and the outer layers of the atmosphere will be puffed off in a series of concentric shells, forming a glowing planetary nebula. Only the core of the sun will remain, as a slowly cooling, dense, white dwarf star. And that will be the long, slow death of the sun.

You give the **sun** another **5,000** million years before anything **changes.** When do you think the **world** will end?

Once the sun gets to the red giant stage just described, which will be in about 5 billion years, then that's when we can confidently say the Earth has had it. Every star has a certain lifetime and at the end of that lifetime, when it's run out of fuel, it dies. Different stars die in different ways – some explode, some become black holes and some become red giants and die gradually. A red giant is a huge star which isn't very warm so it's red rather than bright yellow or white (a bit like a poker heated to yellow heat in the fire and allowed to cool). When the sun becomes a red giant it will swell up so much that Mercury and Venus will be swallowed up inside it, and the poor old Earth will be orbiting just a few million miles from its surface. This will burn off the Earth's atmosphere, heating it to the extent that nothing will be able to live.

Is there anybody **else** out **there?**

If you mean intelligent life, probably not. There's certainly no evidence. However, any astronomer would have to be very brave to say that the Earth was the only life-bearing planet in the universe, and there are plenty of people who'd argue that intelligent life is thinly scattered in our own galaxy.

So, if we are talking about life which is similar to ours, what does it need in order to exist? For a start, it needs a long period of stability to evolve beyond microbes into complex animals and plants. So the first requirement would have to be a stable sun. This immediately rules out 90 per cent of the galaxy's 200 billion stars – they are either too cold and feeble, or too hot and short-lived.

Another essential for life is the presence of liquid – most likely water, which must be in a liquid state in which molecules can mix because in order to form complex molecules, chemicals need to be able to mix together well. This puts an even tighter limit on the requirements for life, for although water molecules are widespread throughout the universe, water is only liquid in a narrow range of temperatures and pressures (0–100°C under Earth's atmospheric pressure). So for liquid water to survive on a planet, it would need a substantial atmosphere, and a stable orbit around its star at roughly the same distance as the Earth. This is why there is no life on Mars or Venus – they are either a little too hot or too cold.

Those two requirements alone rule out any other solar systems we know about, but bear in mind that smaller systems are hard to detect and might well be out there. As far as we can tell, Earth is a unique planet, in a perfect position for our kind of life to flourish. The chances of there being anywhere similar must be very small indeed.

2

Cats, Dogs and Animals in the Wild

The Chicken, the Egg and Swimming Kangaroos

Which came **first,** the **chicken** or the **egg?**

If you're expecting me to say there's no answer to that question, you'd be wrong. The egg came first.

Most scientists think that all life on Earth evolved. Evolution is the gradual development of life to suit its surroundings. For example, a worm lives underground so it doesn't need very good eyes because there isn't much to see. So, any eyes worms once had have been lost from generation to generation. You can't change while you're alive, but your offspring can. As things gradually evolve,

they can change quite a bit. Have you ever seen pictures of what we think our ancestors looked like? Big foreheads, lots of hair, longer arms, slouched backs, etc. We weren't really human then; we've gradually evolved to be as we are now.

The same has happened with the chicken. If you go back in history, what we today call a chicken would have looked different. For example, it might have had webbed feet which made it hard to walk. Then, one day, one of the chickens laid an egg and from that egg came a bird that didn't have webbed feet – it was just like our present-day chickens. It had evolved.

But it all had to start with the egg, so the egg came first.

I've been **told**
that it is **possible**
to lead a **COW**
up stairs,
but not down stairs.
Is this **true?**

Yes, it is true. Because of the arrangement of the cow's knee bones, the joint will flex when walking up stairs, but not if you try to lead it down again.

While we're in the farmyard, you might like to know that a full-grown bear can run as fast as a horse; a horse cannot vomit; emus cannot walk backwards.

Is it true
that the **elephant**
is the **only** animal
that has **four** knees?

No. The elephant is the only animal with *eight* knees because elephants have two 'kneecaps' on each leg. But that's not the entire story, because it all depends on what you define as a 'knee'. The Collins dictionary has 'The joint of the human leg connecting tibia and fibula with the femur' and 'A corresponding or similar part in other vertebrates.' So do 'knees' correspond only to the joints on the back legs of, say, a horse and not the ones on the front? In which case all vertebrate quadrupeds, including elephants, have only two knees.

Or do 'knees' correspond to the joints on both front and back legs of such animals, in which case all vertebrate quadrupeds – not just elephants – have four knees? The problem obviously comes from projecting a term coined for a two-legged animal on to four-legged ones. What is slightly more interesting is that elephants are unique in that they are the only animal in which all four leg joints ('knees' if you like) bend in the same direction.

Do **penguins** have **kneecaps?**

Yes, they do. In fact, their skeleton is pretty similar to ours but they keep their knees tucked up under their fur, which is why you'll never find a penguin with cold knees. People think penguins have only one straight leg bone, but in fact their knees are much closer to their hips than ours.

What **age** do **penguins** live to?

The oldest known penguins are about twenty years old. Most penguins don't get that old, however. Less than half of all chicks live through their first year of life and about 90 per cent of the adult penguins survive from one year to the next. The average age of a penguin is probably around six or seven years old.

Why are penguins black and white?

Penguins are black and white to ensure their own survival. When penguins are swimming through the sea, their dark-coloured backs make it very difficult for seals and other predators to see them. Seen from below, their white fronts make it difficult for them to be seen against the light of the sky, making it difficult for seals, sharks and killer whales to hunt them. They can also use their colouring as a way of controlling their temperature. If they're hot, they turn their white bellies to the sun to reflect its heat. If they're cold, they turn their black backs to the sun to soak up its warmth.

Why do penguins walk in single file?

Probably for the same reason that you and I would follow each other if we were walking in the snow. The first one in line compacts the snow and makes it easier for the others to follow, and you can tread with confidence because if the one ahead hasn't disappeared into the ice, the chances are that you won't either. There might also be an element of wind protection, except of course for the poor penguin at the front.

Can penguins get up if they fall on their backs?

Easily! One of the advantages of being plump and round is that it's quite easy for them to roll over from their backs to their fronts, and then it's easy for them to get up. In fact, lying on their front is a comfortable position for a penguin as this is its natural tobogganing position.

Who invented dinosaur names?

The word 'dinosaur' actually means 'terrible lizard', but the names of the individual species, usually in Greek or Latin, often came from either the scientists who discovered them, or some feature which made them distinctive. For example, Baryonix Walkeri means 'Walker's heavy claw': this dinosaur, discovered by Bill Walker, had a big claw. Velociraptor means 'speedy hunter' and Tyrannosaurus rex means 'king of the reptiles'.

How did the **dinosaurs die** out?

We're not entirely sure, but the dinosaurs died out at the end of the Cretaceous period of history along with lots of plant life and sea reptiles. Only the amphibians and mammals survived, mostly untouched.

Why is a tricky question. One theory that's popular at the moment is the asteroid theory. Two American scientists, Walter and Lus Alvarez, suggested that the impact of an asteroid (or meteorite) on the Earth blasted huge quantities of rock debris up into the atmosphere which cloaked the Earth in darkness for several months, or longer. With no sunlight able to penetrate this cloud layer, photosynthesis ceased and plants died out – thus disrupting the food chain. With no plants to eat, some of the smaller animals became extinct, and with no animals to eat, some of the larger animals died out. This is only one theory. There is also an idea that lots of volcanic eruptions, or perhaps one large one, could have produced the same environmental effects as an asteroid impact.

Before we can decide which might be closer to the truth, we must choose between the theories of the gradual extinction of dinosaurs, and the more catastrophic, instantaneous extinction that might have occurred as the result of an asteroid impact. This is difficult because looking so far back in time is fraught with problems. Even a large enough event to cause the extinction of vast numbers of animals needn't have left that big a mark in the Earth's crust. So how and why the dinosaurs and other animals became extinct doesn't have a simple answer, but there are lots of theories.

Can you **really** make a new **dinosaur** from old **DNA?**

Ah, the *Jurassic Park* question. In the real world no dinosaur DNA has ever been extracted. There was some suggestion that some had been found in the past but it turned out to be contamination.

After 66 million years, which is how long the dinosaurs have been extinct, any DNA that might be found would probably be extremely degraded, and to be able to produce a healthy organism you have to have all the genes in its genome. Genomes for advanced creatures tend to be in the order of billions of base-pairs, and the chance of extracting more than a few tens or hundreds of bases from any very old DNA that remains is just about zero. Even if we manage to find lots of DNA there is a large chance that most of it will be junk (in higher animals about 90 per cent plus of the genome is non-coding DNA). So there isn't really any chance of being able to bring dinosaurs back to life.

In the film, *Jurassic Park*, the DNA was transmitted via a blood-sucking insect which had become trapped in amber. It was a clever bit of invention, but the molecules of DNA which carry the blueprint of all life forms are immensely long and complicated. The chances of being able to find even a few broken fragments of the DNA of animals that died and became fossilized more than 66 million years ago is very remote.

How intelligent were dinosaurs?

To get an idea of dinosaurs' brain power, Dr James Hopson from Chicago set about measuring their brain cavities, allowing for the gaps around the outside and for various other factors. He then compared the size of the brain with other animals, and it emerged that the majority of dinosaurs had the brain power expected of an average reptile. So they were neither overly bright nor particularly stupid.

The Stegosaurus, though, had a walnut-sized brain, and might therefore have been extremely stupid. There were, however, some dinosaurs, particularly the smaller, highly active predators, which seem to have had brains that were larger than might be expected, which may well fit in with the way of life of a resourceful and active predator.

Are there any **dinosaurs** living **today?**

Some dinosaurs do live today – we call them birds. They are properly called 'avian dinosaurs' – dinosaurs with feathers. At least, that's what the popular theory says. In 1916 the Danish medical doctor Gerhard Heilmann published *The Origin Of Birds*, after finding many similarities between birds and the skeletons of meat-eating dinosaurs.

In the 1960s a researcher at Yale University found twenty-two shared features and also noted that these were found in no other animals. This is reckoned to be good evidence of the relationship.

Did dinosaurs and prehistoric humans live **together?**

The very last dinosaurs lived on Earth 66 million years ago, and the earliest remains of human beings can be traced back about 200,000 years, so there's quite a gap.

If we stretch the definition of humans to include even the earliest human-like apes which have been discovered in Africa, then it is possible to take the origins of humans back to about 3.5 million years. So there's at least 62 million years between the disappearance of the dinosaurs and the earliest appearance of man.

What, if anything, do we **know** about dinosaur **poo?** I imagine the **world** covered with **haystack**-size piles of it.

Yes, I suppose dinosaur poo has the potential to be of impressive proportions. In fact, some of it has been preserved as fossils called coprolites. Of course, because of the original soft nature of the poo, coprolites are quite rare – certainly rarer than dinosaur skeletal fossils.

Fossilized dinosaur poo is actually quite useful stuff when it comes to figuring out their behaviour. For example, you can tell if the dinosaur was a plant or a meat eater, or both, by careful examination. Coprolite preservation is dependent on its original organic content, its water content, where it was deposited, and its method of burial. For example, coprolites produced by meat-eating dinosaurs were more likely to be preserved than those of herbivores because of the high mineral content provided by the bone material of the consumed prey animals. Also affecting the preservation would be the location in which they were left: a good place would on a floodplain associated with rivers where the poo dehydrated slightly before rapid burial by a river flood.

Most of the coprolites we know of are from Sauropods, which were the largest of the dinosaurs. They walked on four legs and had a very long neck and tail.

Can **animals** commit **suicide?**

Some people say that dolphins in captivity have done so because one was seen to have battered itself to death soon after its companion had done the same. In nature, giving up your life for the good of others is quite common. Bees will die to protect their hive; lionesses can die defending their cubs; some spiders may be devoured by their young; common octopi look after their young and don't feed themselves, sometimes to the point of death. Certain male bees are ripped open by the queen during mating, and male preying mantises may be eaten by their mate during copulation. There are some parasites that seem to cause suicide in their hosts: some bumblebee parasites can make the bee dive into ponds. Some freshwater shrimp parasites make the shrimp swim at the surface until it is eaten.

Why do **dogs** see in **black** and **white?**

It's a common belief, but it's not true. Dogs have colour vision, but it is very similar to that of humans who have red/green colour blindness. Dogs possess only two of the three types of cones – the colour-sensing cells in the retina – which detect blue and yellow (the yellow also detects red). They don't have green. So they can't detect the difference between red and green, but they can see the difference between yellow and blue.

However, dogs' eyes are very sensitive to changes in movement because they contain a greater number of rods (important for black and white vision), which is important for their hunting ability.

Why do **dogs wag** their tails when they're **happy?**

It may not be anything to do with happiness at all. Most dogs wag their tails to indicate a certain hesitancy when meeting new dogs or entering new territory. It's not exactly a sign of submission, but an indication that they aren't looking for trouble – they've just come to say hello.

But, as you will have noticed, there are two types of wag – the hesitant one and the submissive one. A dog will approach with the hesitant one and then, once a pecking order has been established, replace that with a more friendly one.

Why are **dogs'** noses **wet?**

Because dogs can't sweat. Instead, they drain fluid out through their nose, which evaporates, and this is what makes the nose wet. They cool down by panting, which causes more evaporation from the nose, and that way they lose heat.

But there's another reason why dogs' noses are wet, and it's to do with smell. Dogs have an exceptionally good sense of smell and the damp nose acts as a large wet surface that collects scent particles more easily.

Do **fish**
sleep?

Yes, it does seem as though some fish do sleep. It's not a sort of 'going to bed in pyjamas with the lights off' kind of snooze, but more a period in which they could be said to be 'quiescent'. Of course, they don't close their eyes when they sleep because they don't have any eyelids.

Some fish are well organized about their 'sleep': some tropical parrot-fishes, for example, exude a jelly-like substance which, when it comes into contact with sea water, expands and surrounds the fish to protect it while it is quiescent or 'asleep'.

But for fast movers like tuna sleep can be a big problem. It is the force of their forward motion which pushes air through their gills, and so fish like these can never really stop, but they can slow down.

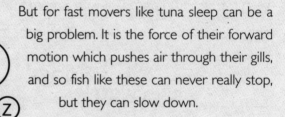

Do **fish**
hear?

Yes, although they don't have ears on the sides of their heads in the way we do for the simple reason that they don't need them. Remember, water is a far better conductor of sound than air, and so the sound simply passes straight into their heads. Apparently goldfish have terrific hearing because their bone structure allows even better transmission of the sound vibrations to their ears.

Because water carries sound very well, many species of aquatic animal communicate by sound. Some people living in permanent house-boats in California heard a buzzing at certain times of the year. Rumour put it down to aliens, but it was eventually found to be the noise of a male toadfish trying to attract females.

I have heard that **fish** don't have any **feelings** and so don't feel **pain.** Is this **true?**

The way that humans feel pain is through receptors in our skin that respond to mechanical, thermal or chemical stimuli. Fish also have these receptors, but that doesn't mean they feel pain in the same way.

In humans the receptor transmits the message via a neural pathway to the higher centres of the brain where we recognize it as that emotional experience we call pain. The brain of a fish is not as well developed, and does not have the same area that perceives pain. So instead of creating a sensation of pain when the receptors are triggered, they probably result in a reflex action without the fish being mentally aware of why.

Do fish **vomit?** **I mean** in the same way that **humans** are sick.

Yes, fish do vomit. In normal eating the muscles of the oesophagus contract in a process known as peristalsis, and these contractions generally force food down to the stomach. But when the contractions occur in the opposite direction, as when swallowing, then regurgitation occurs. The reason ruminants regurgitate regularly is so that food can be chewed more thoroughly. Sometimes fish expel non-digestible food particles and can even expel food when they are excited. Those in the business of selling fish as pets know that it is best not to feed them before transporting them because will they tend to throw up more than usual – and you might think you've bought a sick fish when in fact it's simply excited about the journey.

When our **garden** pond freezes in winter, **how** do the fish survive in the **ice?**

Simple – they find somewhere warmer, and stay there. Water gets heavier as it gets colder, because its density increases, so as the pond cools this 'heavier' water starts to sink to the bottom and the warmer, less dense water rises. Then, when the water cools to below 4°C, something odd happens: the density starts to decrease again which allows the really cold stuff to rise leaving the slightly warmer water at the bottom. Any ice, which is the lightest of all, stays at the top. Although it's still pretty chilly, down in the slightly warmer water is where the fish stay.

Do **fish** get **arthritis?**

No, because fish don't have any ball and socket joints and are continually supported in the water, which removes the strain from their other joints.

Do animals
play?

It depends what you mean by play. When kittens 'toy' with knitting-wool, are they playing or are they practising catching mice? When young foxes 'play-fight', are they doing it for 'fun', or are they getting into training for adult fights? When human adults play the card game bridge, are they just having an entertaining evening or developing their short-term memory? So you can see there is a grey area between the processes of learning and of play.

Perhaps a good definition of play would be 'a complex but seemingly meaningless behaviour that may have a role in learning'. From this, you would expect most playful behaviour to be seen in young animals as they practise for 'the real thing', which is the case. Humans are unusual in the extent to which adults play; perhaps we've got more to learn!

What about non-mammals? Some, but not many, do play. Birds in the crow family, such as choughs, ravens and jackdaws, are considered 'playful' because they are very acrobatic. Choughs, for example, dart up into the air, close their wings and flip over on their backs.

How do **cats** lose heat **if they** don't **sweat** or pant?

By being clever about how they organize their lives. They seek out cool places, think ahead and don't exert themselves to avoid becoming too hot. I'm sure you will have noticed the idle way in which cats behave.

If they do become too hot, they pant with their mouth open, but it's not common so you might not have spotted it. They also choose carefully how they sit or lie, adopting a posture which exposes their maximum surface to coolness, and minimum to heat.

They *do* sweat, through their paws, and a frightened cat leaves behind it damp footprints. They also lick themselves a lot because when the saliva evaporates it cools the cat in the same way as sweating.

What do **cats** see when they look in the **mirror?**

They see pretty much what we do. Their eyes are similar to ours so there's no reason to expect them to see anything other than a reflection. But how they interpret what they see is up for discussion.

We don't think they recognize the image as a reflection of themselves, which is why an animal that sees itself in a mirror or window will approach its reflection as if it were another animal. Cats tend to approach and touch the nose of the reflection, confused by the resulting movement of the reflection, but they never seem to work out that it is their own. Of course, the response of a cat to its reflection is probably the same as that of a child who hasn't seen a reflection before. The difference between a cat and a child is that the child will learn what the reflection is, a cat isn't likely to.

Do cats see in colour?

Cats do have a certain amount of colour vision, but it is not as distinct as in humans. Cats appear to see blue and green but not very much red, and these colours are muted or wishy-washy, just as we would see at dawn or dusk. However, cats are hunters and are therefore far better adapted to movement and low-light vision.

There are two types of receptors in the eye – rods and cones. Cones deal with colour vision and respond to blue, green or red wavelengths of light. Rods are sensitive to light and dark, and are therefore more like motion detectors. Cats have exceptionally sensitive rods, which allow them to detect the smallest changes in the pattern of light and dark. This leads to their reputation for being able to 'see in the dark', which is partly true.

Is it true that cats always land on their feet, and why?

Cats don't always land on their feet, but they usually do. Amazingly, they have no fear of heights and so if they leap into the unknown, when chasing a bird for example, they can often find themselves falling from great heights. If they're falling a short distance, they can usually orientate themselves to land on their feet. But cats falling from great heights can be severely injured.

If you were to watch a falling cat in slow motion, this is what you would see: first, the cat quickly works out which way is up, and then moves its head so it is the right side up. Then it brings its front legs up till they are almost in front of its face to give it some protection. Then there's a twist of the spine so that the front half of its body is in line with its head, followed by a bending of the hind leg in readiness for touchdown. This also brings the rear half of its body in line with the front. In most cases this leads to a soft and upright landing.

The cat's unique skeleton also helps. Their backbones are far more mobile than ours and together with free movement of their front legs this allows them to quickly arrange them-selves into any shape they wish.

Research in America on falling cats has shown that as height increases, so does the injury to the falling cat. But above a certain height – seven storeys – injury rates decline which suggests that providing a cat has enough time, it will get itself sorted out for a landing which causes the minimum damage.

If you gave a **cow** only **grass** to eat, it would **still** put on weight. Where does all that **protein** come from to **make** the muscles?

Plant material contains protein, it's just not as concentrated as in meat. This is the reason why herbivores, such as cows, need to eat such large quantities of vegetation. For example, to produce 20kg of beef protein a cow would need to eat the grass of one hectare of farmland. Elephants are also herbivores, and you don't see many slim elephants. They spend about eighteen hours a day feeding and a typical adult elephant will consume 75–150kg of vegetable matter every day.

If a cow eats **green** grass, **why** is its **milk** white?

The colour of the food animals eat doesn't really determine the colour of what comes out of any end! Remember that a cow has no fewer than four stomachs (the rumen, reticulum, amasum and abomasum) which ensures the almost complete breakdown of the grass. When you break something down into molecules, it no longer has any colour.

So, the real question is, why is milk white? Milk is an emulsion which contains fat, a high protein called casein, complex calcium compounds, and vitamins. None of these is white, however. The white appearance of the milk comes from the scattering of light by the particles in the emulsion. Since, in the case of milk, all wavelengths of light are scattered, and none absorbed, milk appears to be white.

Is it true that all **polar** bears are **left-handed?**

Funnily enough, people who live in the arctic regions where polar bears are found will tell you that this is the case. But there's no proof. However, throughout the world bears are associated with left-handedness, although it seems to be for cultural rather than scientific reasons. In the traditional culture of Vancouver Island, Canada, for example, bear hunters eat with their left hands in order to identify with their prey, since bears are believed to reach for bait with their left paws.

(There is also a rumour that parrots are left-footed!)

I've heard that the polar **bear** is the **only** animal that will actively **stalk** and hunt **humans**. True?

The statistics are against it. The town of Churchill, Manitoba, said to be the 'Polar Bear Capital of the World', was established in 1771, and since that time only two townspeople have ever been killed by polar bears. In fact, polar bears may be a touch cowardly in certain situations: when a bear wandered into a Churchill social club the startled steward shouted, 'You're not a member! Get out!' The bear did. In all of Canada only six people have been killed by polar bears in the past twenty-five years, and in Alaska during the same time period only one person was killed.

In all instances in which a person was killed by a polar bear, the animal in question had been provoked.

Can **kangaroos** **swim?**

Kangaroos do swim. There are parks in Australia where you can see them, especially on hot days. When kangaroos swim they move each hind leg independently of the other. This is quite unusual as they never do so when they are on the ground. When they hop they always keep their legs together!

In case you were wondering, kangaroos that are carrying young in their pouch also swim. To keep the baby safe and dry, the mother kangaroo tightens the muscles around her pouch to seal it.

Can **frogs** hear **underwater?**

Frogs don't have outside ears like we do, but they do have good hearing. They use a thin eardrum called a 'tympanum membrane', located just behind their eyes. In addition to this they have an inner ear, and most frogs have a middle ear.

Frogs will be able to hear underwater, just like us. Sound travels better in water than air, and although it's not common frogs do call underwater to communicate with each other. In fact, there is a frog that is silent on the surface and only calls underwater. It's believed that they do this to avoid detection by predators.

Frog calls can be extremely loud. The forested areas of Puerto Rico are dense with male coqui frogs – there is said to be one in every 10 square metres – and each male stridently calls at the top of his voice to drown out the others, hoping to attract a distant female. The call is so loud that if you wander within a half a metre of one of the little creatures, you will hear the croak at somewhere near the pain threshold of between 90 and 95 decibels. This is almost as loud as a jackhammer (100dB).

Why do animals have tails?

There's no overall answer. Different animals use their tails for different purposes. Kangaroos use their large tail for balance when hopping or at rest, the tail acting as the third leg of a 'tripod'. This is rather like using the tail as an extra limb, something which monkeys do when using their tails to hang from a tree.

Rodents also have long tails which help them balance, while squirrels can also use their tails as shelter. Sea-horses' tails are their only 'limb' and they curl them around reed stems in the water to keep themselves anchored.

Birds' tails have the dual function of providing crucial balance and control during flight, and in some species the male bird uses his tail as a display to attract females – an excellent example being the peacock's tail.

Fish, sharks and dolphins, etc., use their tails to help propel them through the water, and so do tadpoles which lose their tails as they grow into adult frogs/toads and adopt a more land-based lifestyle.

Cows' tails help them remove flies and debris from their rear end, as do horses' tails. So it would seem that these tails are for keeping the animal comfortable and groomed.

Tails can also be used to transmit information: when a rabbit is startled and breaks into a run, the white underside of its tail bobs up and down alerting other rabbits to a potential danger.

In domestic animals like cats and dogs, people feel that the position of the animal's tail tells them a lot about how the animal is feeling. A dog wagging its tail is happy, and when cats are being especially affectionate (or seeking food!) they stick their tails up and purr for attention.

There are almost as many reasons why animals have tails as there are animals which have tails!

Would it be **possible** to **run** across the backs of **alligators** like they do in the **movies?**

Yes, but you'd have to persuade the alligator to stay still, otherwise it would be like trying to run across floating logs. But there's no reason why an alligator's back couldn't support your weight. The other problem is that when an alligator decides to breathe out, it sinks. So as well as lying still you'd have to persuade them to hold their breath.

If **one** of them gave chase, how **fast** could an **alligator** run?

They can go really fast if they gallop. The fastest ever recorded reached speeds of up to 17kph (10.6mph). In general, they can reach 14kph, and this can be beaten by an average human, especially as alligators can sprint only short distances. But they wouldn't normally give chase; they work by cunning and will lie in wait for their victims. They have a remarkable talent for acceleration, and can pounce on a victim before the poor thing has even had time to react, let alone try to escape.

A word of warning. Don't think you can get out of an alligator's way by climbing a tree – patience is one of their virtues and they will sit underneath with their mouth open for a week if necessary.

Have **dogs** got a better **sense** of smell than us?

A dog's nose has four times the volume of ours, and while a human nose has about 5 million ethmoidal or olfactory cells, some dogs have over 200 million. A dog's nose is specifically designed for scent detection: they are large and wet, which helps collect and dissolve scent particles, and their noses can do it far more easily than ours. When a dog finds a scent it starts salivating; this is also part of the scenting process because the wet tongue helps to pick up and dissolve even more scent particles.

How can animals eat **raw meat safely** while humans **can't?**

Animals in the wild eat raw meat all the time and have been doing so for thousands of years. Humans *can* eat raw meat, and in some places steak tartare is considered a delicacy. But we usually cook our meat for two main reasons: first, we prefer the taste of it that way and, secondly, in order to protect ourselves.

Animals eating raw meat will usually eat it fresh; they don't have to transport it around and deliver it to shops and restaurants, and this time-factor is important when it comes to contamination. Humans have a very poor tolerance to the many micro-organisms that can be found in meat, and they can make us very ill. And of course, the older the meat is, the more of these dangerous micro-organisms are likely to be present. Cooking the meat will destroy almost all the damaging bacteria and viruses.

Animals have built up a far better tolerance to such contamination. Domestic pets such as dogs and cats come somewhere in between us and their wild relatives, and have slightly different ways of dealing with foods. Cats protect themselves mainly by being very careful feeders, relying on their phenomenal sense of smell to warn them if their food is 'off'. Cats will also eat grass and make themselves sick if they have to. Dogs are scavengers, and will eat pretty much whatever's going as their digestive system is incredibly tough and can cope with almost anything, but they too will vomit readily if they eat noxious things.

But we poor humans are far less well equipped than our pets, so we cook meat because we prefer it that way and it's less dangerous for us to eat.

3

Birds, Bees and Creepy Crawlies

Sneezing Birds and Spiders' Webs

Why don't **birds** **bump into** one another when they're **flying?**

They would if they didn't have such fantastic reaction times. Think about children playing netball. Each player is supposed to mark their opponent so that they don't catch the ball, and they do that by keeping an eye on them and reacting when they change direction or speed. We're not very good at it compared to other animals. In fact, we're quite slow.

Birds' reaction times, however, are much faster. A bird can react within a split second to the bird next to it changing direction. If you have a flock of birds, each keeping a close eye on the bird next to it, the whole flock can appear to change direction instantaneously because they're all reacting so quickly. But if you took a video of a flock of birds flying and slowed it right down, you'd see that it wasn't instant-aneous. There is a delay between one bird moving and the next one copying it, but the reactions are still fast enough to avoid collisions.

Why do the **birds** all sing
first **thing** in the morning?
I love to hear the **dawn chorus**
even if it can be **very** noisy,
but I would **dearly**
like to know **why** they all sing
at the **same** time.

There is more to birdsong than just its musical content. In fact from the bird's point of view, that's the least of it. Birdsong is all about territory and defence. It is used to attract mates and warn off rivals. It can also alert other birds to danger and young birds use it to tell their parents they are hungry.

Without doubt birdsong is best heard at dawn. It happens around the world from rainforest to temperate meadows, but we are not certain why. It could be because dawn is often a still and tranquil time of day, so the sound carries better. Measurements have shown that the song can carry twenty times further than at other times of day. Also, it's a time of day when the birds don't have much else to do: there's less light for hunting, and insects will still be hiding after the cool of the night and so the birds sing. Enjoy! It's one of the wonders of the world.

Do birds **sneeze?**

They certainly do, but it's pet birds you are most likely to catch doing it, and this may be because of human contact. As with humans, sneezing could be a sign of infection.

How can **skylarks hover** in the air, continuously **sing** and **breathe** all at the **same** time?

It's a kind of showing off. By demonstrating they can manage all three things at the same time, the male skylarks are showing themselves to be the best male bird around, and the females should take no notice of the others.

But even the multi-talented skylark can't sing for ever, although it might sound like it. The impression of it comes from our inability to discriminate between a series of rapid acoustic events in time. There are, in fact, brief gaps in the song during which the birds can breathe, though the exact details of how they integrate their breathing and singing remain sketchy.

If I put my head underwater everything looks blurry. So how do diving birds, like ducks, see clearly underwater?

Things look blurry underwater because our eyes cannot focus properly. This is because light travels the same way through water as it does through the cornea of our eye, and doesn't get bent as it usually would when passing from one medium to another. This means the image doesn't focus properly, because it hasn't passed through air first. That's why, when you wear goggles, you restore the air/cornea boundary and then you get a clear image. Fish have much thicker, sharply curved lenses so they can focus in the water.

As for diving birds, there are two possible ways in which their vision works underwater. The first is that the lens in their eyes can be thick or thin to enable adaptation to seeing in or out of water. The second, and most likely explanation, is that the birds are aware that the fish are not where they appear to be, because the image reaching their eye has been bent by refraction, and they compensate for this when they make their dive. This can make ducks look cleverer than they sometimes are.

Can **owls** **really** turn their heads in a **full** circle?

No, their heads can't go all the way round because it would damage their nervous system. But they can turn their heads round to a far larger degree than other animals.

The field of vision in birds ranges from only a few degrees to a full 360 degrees, and this is a good guide to whether they are predators or prey. Prey species tend to have eyes on the sides of their heads which gives them 360 degrees of vision, so they can scan as much of the world as possible to spot approaching danger. Predatory species tend to have eyes nearer the front of the head which gives them a wide field of binocular vision, giving them a great ability to judge size and distances and see finer details. It also makes the eyes better able to see when the light is poor.

Owls have a 60 degree field of vision towards the front, but an extensive blind area of about 130 degrees behind. The field of vision of most other birds falls within these two extremes. Owls can revolve their heads further to counteract this extensive blind zone.

Why don't **woodpeckers** get a **headache?**

Actually, woodpeckers have very small brains, and the brains they do have are suspended in fluid. Together with shock absorbers in the beak, all that hammering away probably has little effect on their skulls.

Why can't chickens **fly?**

Good question! They've got everything they need to fly, including wings and air sacs around the lungs, while their bones contain numerous air cavities that make them light for flying. But the problem is that for centuries they've been seen as a source of food, and through the process of domestication they've simply forgotten how to do it. Archaeological evidence places the chicken in human settlements in Indo-Pakistan by 3250 BC and since then, by selecting them for the quality of their meat; we've eventually bred a bird that gives us plenty of tasty meat, and at the same time we've bred out of it the ability to fly (although some breeds can make a miserable attempt at it).

You've probably noticed that a chicken tends to have some white meat on its carcass and some dark meat. This is due to the amount of a pigment in the muscle called myoglobin, which is closely related to haemoglobin, an oxygen-carrying component of blood. If the muscle is used repeatedly for extended periods of time, the amount of myoglobin in the tissue increases. That is why migrating birds such as ducks and geese have darker meat in the breast muscles – the muscles used for flight. Chicken's legs are also muscled with darker meat because the birds tend to use them more than those in the breast. The meat on the wings is white, of course, because chickens don't fly any more.

Why do **pigeons'** heads **bob** back and forth when they **walk?**

The pigeons we see in our cities are feral (wild) birds and they greatly fear birds of prey and so will be constantly on the look-out for predators. Having eyes on the sides of their

heads, they already have a wide field of view, but they try to increase this by bobbing their heads back and forth. If it makes them look edgy, it's because they are.

That's one theory, but there are others. It has been suggested that the pigeon's head doesn't move at all, it's the body that's moving beneath it. After German scientists videoed pigeons over a long period of time and examined their movements frame by frame, they decided this was the case, and suggested the motionless head allowed the pigeons to assess space and distance better.

How do **homing** pigeons **find** their way home?

No one has proved it either way, but there are two current theories. The first says they use some kind of 'odour map'

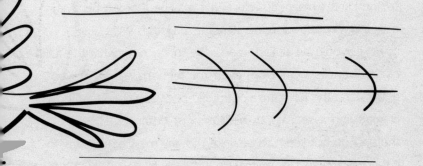

which is imprinted on their minds from an early age in a variety of wind conditions, and once they've got the scent of home they'll always return. The other, more likely sounding theory, is that they use the Earth's magnetic field to fly to a precise latitude and longitude. But no one knows.

I have heard
that the **magnetic** poles
are overdue to **'flip'**.
Would pigeons
still **be able**
to **find** their way home?

Even if it were to happen it would take anywhere between a few thousand to seventy thousand years to complete. From a human point of view, it would hardly matter now that we rely less on magnetic compasses for navigation. But for animals it might be a different story. Certainly it has been shown by experiment that the migration of turtles is strongly affected by the Earth's magnetic field, and it may be that fish behave in a similar way, but our understanding of both is slight.

So, it would be almost impossible to say now what the full consequences would be, even for pigeons, as we do not understand the full nature of the flip or even understand how animals find it so easy to navigate vast distances. The flips in the past do not seem to be linked to any mass extinctions so if there is an effect on animal life it is likely to be small.

How do bumblebees **manage** to fly? They look **far too heavy** for the size of their **wings.**

It's because the bees are obeying different laws of motion to the ones which limit the way in which we fly. Certainly if an aeroplane was the same size and shape as a bee it wouldn't be able to fly. But bees and aeroplanes fly in very different ways.

The motion of air under and over the wings is what keeps an aeroplane in the air: the shape of the wings means that air moves faster over the top of the wing than below it, which causes a drop in pressure above the wing and an increase in pressure below the wing. This gives the aeroplane lift.

Bees fly more like a helicopter. Their wings are in constant motion, and it is this that provides the lift. Because bees are very small, from their point of view the air behaves more like a viscous fluid, such as treacle, and they create vortices down the outside of their wings which help provide lift and forward motion.

Why do **flies** buzz **continuously** around the light bulb, even when the **light** is off?

They don't; it just looks as if they do. This is because you only notice them when they're annoying you. Also, the fact that you are there has probably stirred them up. Otherwise, they'd just sit there and get on with the disgusting business of spreading disease.

However, once in the air it's true that they do seem to like buzzing around the middle of a room, unless it's bright daylight outside when they prefer that. There's a theory that they don't like corners; or it could be that they're planning to use the light fitting as a perch from which to launch an attack on a rival, or for attracting a mate. Female houseflies aiming to perch on a light fitting will be intercepted by the male lucky enough to be patrolling the airspace nearest to it, and so the males compete to occupy this top position from where they can dart out to chase away any other flies which are threatening to invade their airspace.

Here's a good game if you want to annoy the flies as much as they annoy you: the next time you see a housefly circling purposefully beneath one of your ceiling lights, try throwing a dummy fly at it (a fly-sized bit of paper will do fine). The fly will almost certainly break off from its horizontal circling to chase the 'intruder' as the projectile arcs through its airspace.

How do **flies**
land on a **ceiling** and take off?

Scientists used to think they performed a quick roll and hit the surface with either their front legs or their back, and then rotated onto the others. But by filming flies in great detail they have discovered the whole business is far more graceful than they thought. As a fly approaches a ceiling, it moves its front legs up towards the ceiling, and this is the first point of contact. Attached by its front legs, it uses the momentum from its flight to 'flop' the remainder of its body onto the ceiling.

How do **spiders**
move their legs?

Spiders' muscles are attached to the inside of the exoskeleton and work antagonistically, meaning in 'opposite pairs', which is the way ours do. The movement of their legs is also partly hydraulic. Spiders can stretch their legs by raising their blood pressure to such an extent that a jumping spider can create a force that makes it capable of leaping twenty-five times its own length.

Why don't spiders stick to their own webs?

If a spider is taken off its web and thrown back at it, it might well stick to it. But their feet, the tarsi, are covered with a non-stick secretion and in the normal course of spinning a web, this prevents such accidents.

Will a spider ever use another spider's unused web?

Generally not, although there are instances where some spiders will. Some males, for instance, will invade a female's web when courting and after mating the male will hang around the web and nip in for a feed when the female is not looking. If for some reason the female dies, the male continues to use the web until the environment destroys it.

There is one spider, the pirate spider, that *will* creep into another spider's web very slowly so that you can hardly see it move at all. The other spider knows that something is not quite right, but by then it's too late because the pirate spider, when it's close enough, will reach out and bite the other spider on the leg, injecting it with a very toxic venom. The other spider is instantly killed and the pirate spider eats it for dinner.

How do spiders get from place to place when they spin their webs?

They wait for a fair wind! A web is made from silk, which emerges from the spider's abdomen as a liquid which solidifies as it dries in the air and forms very thin threads. It is incredibly strong, stronger than any material or metal known.

Unfortunately spiders can't shoot a web like Spiderman can. Instead, they are to make use of the wind. The spider hangs on a thread until it is caught by a gust of wind and carried to another place where it can attach the other end. Once this first thread is in place the rest of the web is easier to construct.

Why are the **webs** made in different patterns? **Is it** to trap different **kinds** of insect?

Very interesting question! In fact you can tell a spider by its web; it's almost like a fingerprint. Yes, the different webs do target different types of prey; for example, an orb web close to the ground is aiming to catch jumping insects such as grasshoppers, while an orb web above the ground will trap flying insects.

Vertical webs high in the vegetation catch flying prey, while those lower down catch jumping prey. Horizontal webs catch prey insects that fall off the surrounding plants or jump into them. Those set at an angle catch a mixture of everything.

A **tiny** spider often drops
four or five **feet** into our room
in the evening,
hangs **around** for a bit,
then **goes** back up towards
the ceiling.
What happens to the **thread**
of **spider's** web on which
it was hanging?
Does it **wind** it in,
eat it ... or **what?**

The silken thread just hangs around drifting in the air currents, although sometimes you may not be able to see it. Spiders always leave a silken thread behind them which they use as a lifeline if they fall, or as a guide to find their way back to where they came from. These threads are abandoned after being produced and one can often see the scale of this silk production in the early mornings after a heavy dew in pastures where the whole field is covered with a shimmering mass of silk that wafts about in the breeze. Spiders do eat their own silk to recycle the proteins and to ingest any pollen caught on the sticky spiral. This is an important protein supplement for immature spiders.

Can **spiders** see?
I had one **walk**
right **under**
my foot.

Spiders can have either two, three or four pairs of eyes, depending on which family they belong to. You might think this would mean they should have good eyesight, but actually they have very poor vision. So, they use tactile senses to find their way around and seek out prey. They have one set of structures to tell them where bits of their own body are, such as their legs, and others to tell them about their surroundings.

The hairs which cover many spiders' bodies are part of their sensing mechanism. If something touches a hair a nerve connected to that hair lets the spider know that there is something there. They have other more specialized hairs too, called *trichobothria*, which pick up smaller vibrations, such as the buzzing of an insect's wings.

Spiders have one more way to 'see' without using their eyes. They use things called slit sense organs which are often on their legs. Spiders which spin webs use these organs to tell them when they have caught something by the movement of the web.

So the spider probably didn't see your foot with its eyes, and because your foot doesn't buzz like a bee and you weren't caught in its web, it didn't know you were there until it felt your foot with its hairs!

How do **worms** penetrate hard soil during the **summer?**

Earthworms are largely crevice burrowers, so they seek cracks in the soil into which they can squeeze their bodies and move by what we call peristaltic locomotion – this is when a bulge passes backwards along the body acting as a temporary point of attachment as the animal propels itself forwards. If the soil is very rich in food or very compacted they will essentially eat their way along.

During cold or dry weather many species burrow deeper than usual, cease feeding, curl up into a ball and wait for warmer or dryer conditions to return. When we find them inside hard, dry soil we have to remember that when the worms were active the soil was wetter and softer. The walls of worm burrows get compressed as a result of the worms' movements, and are coated with mucus and urine creating a smooth lining, which is more comfortable for the worms than the soil alone would be.

How does a **glow-worm** glow?

Glow-worms and fireflies use a process called bio-luminescence. Their light-producing organs contain a chemical called luciferin which is stored under a transparent cuticle that has an area of very dense tissue underneath – this probably acts as a reflector. To produce the light, the luciferin reacts with oxygen in the presence of the enzyme luciferase. This produces oxyluciferin and energy, which is emitted in the form of light. Later the oxyluciferin is transformed back into luciferin so that the process can be repeated.

Because hardly any energy is generated as heat (unlike a fire or a light bulb), this is one of the most efficient means of generating light. The brightest fireflies only produce light equivalent to about 1/40th of a candle, but the light is emitted at a wavelength to which the human eye is very sensitive, so they may give off enough light to read a book by. Bright fireflies found in China and Japan have been known to be used by poor students for just this purpose.

What is the average **life** expectancy of a **slug?**

Bad news for gardeners: a large slug can live from eight to ten years. The smaller ones live for about six months.

Why do **moths** fly **towards** light?

If I said it was because they were confusing your bedroom light with the moon, you might not believe me, but it would be true. Moths navigate using the light of the moon as a steady reference point and fly in a reasonably straight line keeping the moon on one side. When a bright artificial light is present they try to do the same thing, but to keep it in a fixed position they end up flying round in circles. The brightness of the light disorientates them and their orbits get smaller and smaller until they eventually hit the light.

What did the **clothes** moth eat **before** there were clothes?

Clothes moth larvae, as well as attacking woollen clothing, also live in bird and mammal nests. They feed on a mixture of detritus and the fur/wool of the animal, plus a certain amount of fungal material. So they don't just eat clothing. Clothes moths belong to a small group of closely related species that have developed the almost unique ability to digest keratin, the protein that makes up fur, wool, hair and feathers (as well as toenails and dead skin). Before we started storing suitable food for clothes moths in the form of a winter wardrobe, they thrived in other ways, as they still do.

How far can **ants** see?

It depends on the type of ant. Some worker ants have well-developed eyes and can leap from branch to branch, others

have greatly reduced eyes, and worker army ants have no eyes at all. Some ants must have top-class vision: the 'jumping ants' of India leap up to a metre in the air to catch flying prey with their long mandibles. Although we don't quite know how they do this, they must have pretty good vision. But their 'seeing' is not like ours. We see one big picture, but insects see a lot of little pictures, rather like a shop window full of television sets all showing the same programme.

What is **life** like for an ant?
Do they have time off?

Ant life has four stages – egg, larva, pupa and adult – and it spans a period of eight to ten weeks. The queen spends her entire life laying eggs. The workers are females and do the work of the nest; the larger ones, the soldiers, defend the colony. At certain times of the year many species produce winged males and queens who fly into the air and mate. The male dies soon afterwards and the fertilized queen establishes a new nest.

Whether ants get any time off all depends on the

are only active if the temperature is high enough. So on cold days and nights they will remain dormant in the nest, but as soon as the temperature increases they are up and about. They have a good set of compound eyes which enable them to navigate using the sun. So even in the tropics where it is warm all the time, the ants are only active during the day as it is difficult for them to find their way around at night.

Do **ants** have **blood** and **bones?**

No, they don't have bones. Their skeleton is made up of a chemical called chitin which is a waxy chemical not unlike plastic. Chitin covers the outside of the ant's body, so you could say that ants effectively wear their skeleton on the outside.

Insects do have blood, but they only use it for carrying food around their bodies. Humans use blood to transport oxygen. Ants have a simple heart which pumps blood round the top of their body but it consists of a simple, long, thin tube.

How do **insects** smell?

Insects have large numbers of 'smell organs' consisting of *sensilla* which are small hairs modified to sense either touch, smell, taste, heat or cold. Each *sensillum* consists of just one sense cell and one nerve fibre.

4

Down
to
Earth

Autumn Leaves,
Ripe Tomatoes and
Germs

What are **plant leaves** for?

Think of leaves as large solar panels which trap sunlight for the plant so that it can produce its own food. Without light a plant cannot survive: try keeping one in a dark room and watch it rapidly wilt and die.

Leaves also have tiny holes in the surface, mainly on the bottom side. These holes (called *stomata*) allow air in, and the carbon dioxide in this air is also part of the food-making process. The last ingredient plants need to make food is water, which they get from the soil through the roots.

The reason leaves are so thin is because the carbon dioxide must move through the leaf, which is easier if it has a shorter distance to travel. Also, it provides a larger surface area to trap the sunlight.

Why do **leaves** change **colour** in the autumn?

The reasons for the drastic colour changes of autumn are quite complex. Basically, leaves supply the food that allows a tree to live and grow. Soon after they unfold in the spring, new young leaves begin making food through a complicated process called photosynthesis, which uses the energy of the sun to combine the raw materials taken from the soil and the air. The basic ingredients a plant needs for photosynthesis are sunlight, water and carbon dioxide, which is also the gas we give off when we breathe.

Carbon dioxide is taken into the leaf through tiny openings on its surface. Water is drawn up from the soil by the roots and carried to the leaves through tiny veins. When these raw materials enter a leaf that is exposed to the sun, photosynthesis takes place and the plant makes food for itself. In the leaf are tiny particles containing a green pigment called chlorophyll. This pigment not only gives leaves their green colour, it makes photosynthesis possible.

In the autumn the amount of sunlight decreases and trees stop making food. Because photosynthesis is ending the green pigment is no longer needed, and the leaves destroy it. As the green begins to fade, yellow and orange pigments, which had been hidden by the green, start to appear. Bright sunshine and cool night-time temperatures are needed for the bright reds to appear. In years of early frost, the leaves are more likely to be brown than red.

Why are plants **scented?**

It's all to do with love and romance, sort of. Since plants, unlike most animals, are unable to move from place to place, they have evolved a set of features that in effect allow them to find a mate. By attracting insects and other animals with their flowers, and by directing the behaviour of these animals, cross-pollination occurs. This is the nearest a plant gets to love-making.

In the process of evolution plants learnt that the more times animals and insects were attracted to them, the more often cross-pollination would occur and the more seed they would produce. To make themselves more attractive, plants might produce nectar to feed the insects, coloured petals or scent.

Do **bacteria**
have sex?

There seems to be no fun attached to the way a bacterium reproduces, which is usually by a method called binary fission in which they simply split into two to form two new cells that are identical. No 'partner' is required to help this process along.

However, there are some bacteria who do actually mate. Bacteria have small, hair-like structures called pili on their surface. One bacteria can join pili with another bacteria, forming a continuous tube between the two cells. Small segments of DNA called plasmids can then pass from the donor bacterium to the recipient, so passing on useful genes. However, this does not produce any offspring – it's just a transfer of information, so it can't be properly described as 'having sex'. But any genes that are given to the recipient are passed on down the generations as this bacteria cell divides in the usual asexual way.

The way in which bacteria mate and pass on genetic information is quite important for human health. If a particular strain of bacterium develops, say, a resistance to an antibiotic, then this is the process by which the antibiotic resistance is passed on to other bacteria until we have a strain which cannot be controlled when we get sick.

Is it **true** that there are about **a million germs** on every pin-head?

First, what do we mean by germs? At its simplest, you could say a germ is any living organism which we cannot see and which also makes us ill: it could be a bacterium, a virus or a fungus. It is true that there are probably a million bacteria on the head of a pin. Bacteria are all around us. However, unless we stab the pin deep into an arm or a leg, these bacteria will not harm us.

Do *all* **bacteria** make us sick?

No. In fact surprisingly few of them make us ill at all. There are roughly 10,000 known bacteria, and probably as many again which are unknown. Even so, there are only about thirty bacteria which are dangerous and all these are well known. No new bacterial disease has been found for many years: the new diseases you read about are almost always caused by viruses. Bacteria are with us all the time in our guts, on our skin and inside all the major body openings. These bacteria are actually useful to us because they compete with the few bacteria that try to invade our bodies to cause disease.

If you have ever had to have an antibiotic for, say, a chest infection, you may have noticed that taking the antibiotic gave you diarrhoea. This is because the 'good' bacteria which normally inhabit your intestines are also killed off, and as these organisms help to 'bulk out' faeces, the result is very loose and liquid stools. Some people also find that taking antibiotics makes them much more susceptible to thrush, a fungal infection, because when the normal bacteria are killed the fungus can more easily gain a foothold and reproduce. Some people swear that eating live natural yoghurt while taking antibiotics can help to avoid these problems.

How does a **mushroom** breathe?

They don't breathe in the sense that they have lungs, but they do absorb oxygen from the environment to power the metabolic processes of the tissues. The important thing to realize is that a mushroom is not the whole organism – it is only the reproductive part of it. The majority of the fungus is to be found growing within the material on which the mushrooms are found. In the case of a mushroom found on a rotting log, the wood of the log will be permeated by a network of fungal filaments called *hyphae* and the mushroom is formed from these *hyphae* which group together. As they are packed quite tightly, it is not easy for the oxygen to diffuse into all of them from the outside surface of the mushroom, so the fungus gets over this problem by having a sort of mini-circulation within the mushroom stalk. This carries oxygen and other goodies into the middle of the structure to keep all the hyphal filaments going.

How does **water** get from the **roots** of a plant to the **leaves?**

By what is called the transpiration stream. When a molecule of water evaporates from the surface of a plant's leaf, it pulls up molecules of water below to replace it, resulting in a continuous stream of water upwards through the plant. Water molecules naturally stick together and to the sides of the vessels they are being pulled along – this is called cohesion and it is this cohesion between water molecules which maintains the continuous stream. There is also a membrane between the roots of the plant and the stem xylem, which are the vessels that transport the water, and water has to cross this living cell membrane before it can rise to the leaves.

Why does a **cactus** have **such a fat** skin?

The cactus is basically a big fat stem covered by an incredibly thick waxy cuticle. Because the cactus is naturally found in an extremely dry habitat, conservation of water is of the utmost importance. This is the reason why a cactus has no leaves – it would lose water too easily. Instead, the cactus uses its stem to perform the same function as the leaves, which is to absorb sunlight, perform photosynthesis and make food. The very thickness of the stem reduces water loss, as does the fat, waxy cuticle or skin.

What makes a **nettle** sting?

There are tiny hairs on a nettle leaf that look like needles and can easily penetrate your skin. At the base of each needle is a bulb filled with formic acid, which enters your skin along with the needle. This causes an allergic reaction in your skin making it red and itchy.

Do **plants** feel pain?

You've first of all got to decide what you mean by pain, and there is philosophy here as well as science. But let's say pain is 'a response to physical stress aimed at reducing that stress'. Research has shown that plants do have a stress response. When a leaf is cut, it releases a gas called ethylene from its surface. This is a kind of pain response: the release of ethylene is a signal to the plant to take measures to withstand stress. That would fit in with our definition of pain. So, on that basis, plants do feel pain.

But if you're going to use this simple definition of pain, then you must argue that anything living feels pain, because all organisms have stress responses. Bacteria have many for example, and their response to heat has been particularly well studied. So can we assume that bacteria feel pain?

At a very simple level, plants have systems and responses that look like pain. But this is where the philosophy comes in, because pain means much more than a simple chemical response. So, perhaps you could say that plants do feel pain, but not in the way you and I do.

Why do we need **plants?**

Without plants we would not exist. All the energy which allows us to live has come from the sun, but humans and other animals can't utilize that energy directly. We have to rely on other organisms, or life forms, to do this for us. Then, by eating these organisms, the energy is passed on up the food chain to us.

The process by which energy from sunlight is captured in living organisms is called photosynthesis. There are roughly half a million kinds of organism that can photosynthesize, and they are all plants, algae and some types of bacteria. These organisms turn sunlight into the molecules which we need to survive, and the only way we can get these important molecules is by either eating plants or by eating another animal that has eaten the plants.

Plants are also important to us because they release oxygen when they photosynthesize, and this oxygen is necessary for the survival of nearly all organisms, including the plants themselves.

Remember, the only reason human beings and other animals exist is because plants were here before us making the world suitable for us.

Do plants
sleep?

If you think of sleep as a period of inactivity (rather than a change in consciousness as in human sleep) then you could say that, yes, plants *do* sleep.

Many plants have a daily cycle or rhythm. Daisies open their petals in the day and close them at night, and botanists refer to these as 'sleep movements'. One possible reason for this behaviour could be sensitivity to different wavelengths of light.

Plants can certainly tell when it's day or night, and how long the darkness lasts. They contain a pigment called phytochrome which exists in two forms: one is sensitive to the red light that the plant receives in the day, and the other to the far-red light more abundant at night. The relative amounts of the different forms of phytochrome enable the plant to tell day from night. Interrupting a plant at night with a burst of daylight can disrupt its functioning, which is why some plants fold up at night to reduce the possibility of this happening.

Why **can't** you melt **wood?**

A liquid is a collection of mobile molecules – in other words, they move around easily. But wood is made up of lots of cellulose, which consists of very long chains of polymers. Now, long chains of things can't move around very easily. Also there are hydrogen bonds between the hydroxyl groups in the polymers that hold everything together. In other words, you'd have to put so much energy into breaking these bonds that before the wood melted it would decompose and no longer be wood.

Someone told me that glass isn't a **solid**, it's a **liquid**. **Were** they joking?

No, glass is not a liquid: it is as solid as any other solid. Try hitting yourself over the head with a glass bottle and you will find it to be extremely solid! However, what they may be getting at is the fact that glass is what is known as a super-cooled liquid.

For all solid elements, and many solid compounds and mixtures, heating causes them to melt, and this melting occurs at a very well-defined temperature. Just a fraction of a degree below the melting temperature and the material is solid and will retain a well-defined shape, but just a fraction above and the material is liquid and can be poured.

This is not true of glass. Glass becomes softer and softer the more it is heated, and has no easily defined melting or freezing temperature. Toffee is the same. Try eating a toffee which has been kept in the fridge or freezer for a short while and you will find it amazingly hard. If you heat it you will find no well-defined melting temperature, it just gets softer and softer. It is possible to define a temperature for glassy materials, called the glass temperature, which gives an indication of the temperature range around which the increase in viscosity on cooling is most rapid. But it's not the same as a freezing or melting point.

Where do we get the helium for putting in balloons?

Helium is the second lightest element – only hydrogen is lighter. It's a colourless, odourless and tasteless gas and is used for many things, including filling balloons! It's also used to pressurize the fuel tanks of rockets, as a coolant, and in high-pressure breathing operations underwater.

Helium was discovered in the sun's atmosphere by the French astronomer Pierre Janssen in 1868. Then in 1895 the British chemist Sir William Ramsay discovered helium in the Earth's atmosphere, but in very small quantities. It's also found in radioactive minerals and mineral springs. But these are only small sources, not enough to provide the quantities needed for vital balloon filling, given the number of birthday parties that are happening in the world.

Luckily large volumes of helium are found in natural gas reserves in the United States, and smaller supplies have been discovered in Canada, South Africa and the Sahara Desert. The helium is isolated from the natural gas by liquefying the other components at low temperatures and under high pressure, which leaves a mixture of gases containing just over 90 per cent helium. By passing this over cooled, activated charcoal, the other gases in the mixture are adsorbed, leaving essentially pure helium.

Why doesn't iron dissolve in water?

All the particles that make up a solid are stuck or bonded together. These bonds can be weak or they can be strong. To dissolve something, the bonds between the particles have to be broken.

If something is a solid, then the particles are all quite happy as they are, sitting bonded together, and to persuade the particles to separate you have to present them with something more attractive. So if you have a liquid into which you want to dissolve a solid, the liquid particles must be able offer good bonding interactions with the individual solid particles. Then the solid particles will separate from one another and form lots of bonds with the liquid particles, and will be quite happy with their new-found friends.

In general, things are likely to dissolve in similar things, because there will be similar bonding opportunities between the solid and liquid particles. But iron and water are very different materials. Water is good at dissolving lots of things, but not much good with metals. In a metal all the particles sit quite chummily together, and water can't offer anything as attractive.

When you hold an **egg** by the **pointy** bit and try to crush it you can't. But **if you** put the same amount of **pressure** across its waist, it would soon **smash** to pieces. Why?

An egg is a remarkably engineered structure. It is true that if you hit the side of an egg with a spoon it will crack, but that is because the shell is thinnest there and easily damaged. But because of its shape, pressure applied to the pointed end of the egg causes it to behave as if it were a structural arch in a building, or a bridge. In arches, the loads compress the whole structure, and under compression the calcium carbonate of which eggshells are made is very strong.

Is it the **skin** or the actual fruit
of a banana that **produces** ethylene?
And **why** does the fruit
change from a well-camouflaged
green to a **bright** yellow?

Ethylene (also called ethane) is a plant-ripening hormone and it is produced by the whole fruit, not just the skin. It is produced by any cell in the banana in which membrane lipids are oxidized to produce unsaturated fatty acids.

As regards the change of colour: when ethylene is produced it causes the breakdown of fibres in the fruit, making it soft. It also causes the breakdown of starch to make sugars, making the fruit sweet, as well as the breakdown of chlorophyll, causing the green colour to disappear. The pigments that make a ripe banana yellow are there in green fruit too, but are hidden by the chlorophyll until the breakdown starts to occur.

If I want a tomato to **ripen**,
will it **happen** faster if I put it
in a **sunny** place,
rather than in
a **dark** cupboard?

If you are hoping to get some flavour into a dreary supermarket tomato by ripening it, you may be wasting your time.

No matter what you do, you will never get it to develop a taste like a tomato that has been allowed to ripen on its vine.

Commercially grown tomatoes are carefully bred and selected to have a firm flesh, and if you believe that your tomato is too firm and has no aroma but it is red in colour, the chances are that this is as good as it will ever get.

Generally, tomatoes will ripen and develop a bit more flavour if left at room temperature for a few days. They should be stored away from direct sunlight since this softens them without ripening, and strips them of their vitamins A and C. Tomatoes that are stored in the fridge and below 50°F (10°C) will lose their aroma and flavour quicker than those stored at higher temperatures. Apparently, according to food and cooking experts, the top of the fridge is a good place to keep your tomatoes as this is quite a warm surface.

There is a trick that you can use if you want tomatoes to ripen fast. Place them in a paper bag on their own or with a banana or an apple. As tomatoes ripen, they give off a chemical, ethylene, that will stimulate other tomatoes or any other ethylene-releasing fruits and vegetables to ripen. If you keep them in a bag, it means that the ethylene cannot escape and all the fruits will be exposed. As bananas and apples also give off ethylene, placing them in the bag will also speed up the whole process.

Why do eggs **bounce** after being **soaked** in **vinegar** for a while?

Because the egg that has been soaked in vinegar isn't quite the same egg that you put in there in the first place. What

happens when you put an egg in vinegar is that bubbles start forming on the surface of the egg's shell. After seventy-two hours the shell will be gone, and portions of it may be seen floating on the surface of the vinegar. But the egg remains intact because of the thin membrane which is not dissolved by the vinegar.

The shell of the egg is made of calcium carbonate, and when vinegar reacts chemically with it, one of the products is carbon dioxide gas – forming those bubbles seen on the egg. The membrane around the egg doesn't dissolve in vinegar, but becomes more rubbery.

You might also notice the egg has increased in size. This is due to osmosis, which is the movement of the water content of the vinegar through the outer cell membrane into the egg. This movement takes place because the water inside the egg has more materials dissolved in it than does the vinegar, and water will always move through a membrane in the direction of more dissolved materials. That's why the egg gets bigger.

It doesn't matter if you boil the egg before you try this. A boiled egg would be nice and bouncy, but an unboiled egg is more likely to be squishy, like a water balloon.

5

Seeing
isn't
Always
Believing

Mirror, Mirror on
the Wall . . .

I was **looking** in the mirror
and noticed
that **everything** is reversed
from **left** to right.
Why aren't things
upside-down as well?

For a start, you're wrong about things being flipped left and right. If you look in a mirror, the left-hand side of your face is still on the left-hand side and the right-hand side is still on the right. The same is true for tops and bottoms. This is really just a myth that is perpetuated because until you

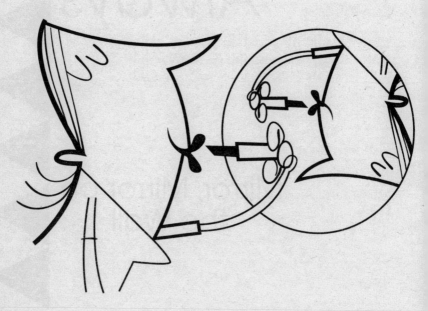

sit down and think about exactly what's happening to the light it actually looks like the rumour is right. There's *no* flipping so there's no reason why things should look upside-down.

I was **wondering** if I could buy
more mirrors
instead of **buying** more light bulbs?
I mean, if I shine a light
on a mirror,
and direct the reflection
into the **room**,
I've doubled the **amount** of light,
haven't I?

A mirror cannot make more light than there already is in the room. You can't make light out of nothing; you have to use energy to create light. You can bounce it around, but that's all. You don't get two footballs by kicking one against a wall and imagining the ball that bounces back at you is some kind of 'new' ball.

Normally, light is absorbed by the surfaces it lands on, which is why black is so dark because it absorbs all the wavelengths of light and appears black: an absence of light. So what happens with the mirror? The mirror simply reflects the light rather than absorbing it. That is why it seems as though there is more light in the room.

I **thought** white surfaces
appeared white because they **reflected**
all the **light** that fell on them.
So if a mirror **reflects**
everything that **falls** on it,
why doesn't a **mirror** look white?

Because a white sheet of paper isn't just reflecting light, like a mirror. White objects appear so because they're absorbing all the colours of light and re-emitting them as one single colour – white. A blue object would absorb all colours but just emit the blue. A mirror is doing no absorbing: it's simply bouncing back at you what you throw at it, so the light undergoes no absorption or re-emission.

I've seen **one-way** mirrors.
They've got them at airports
so the **police** can watch people
as they **pass** through.
But from your side of the **window**
it looks like a **mirror.**
How do you make a **two-way** mirror
so **you can** see through it
one way, but not the other?

You need a piece of darkened glass, which gives it a smoky look, and to this is applied a very thin film of reflective material which is usually made from aluminium alloys. The coating has to be very thin because a certain amount of light needs to pass through it, but because it is reflective too, some of the light falling on it bounces back.

Now, imagine the glass in a wall and that you are a spy. Because there is darkened glass between you and the mirrored surface, your image is not bright enough to pass through, and somebody standing on the far side of the mirror would only see their own reflection whereas you would have a good, if somewhat darkened, view of them. You can turn a one-way mirror into a clear window simply by turning up the light on your side till your image is bright, then you can see through it both ways.

On **our** car, we've got one of those **rear-view mirrors** which you can tilt and it dims the **glare** of the headlights on the car behind. What's happening **there?**

Mirrors are usually silvered on the back and this is where most of the light is reflected. But about 5 per cent is also reflected by the surface of the mirror. On a normal mirror the front and back surfaces are usually parallel and we don't notice the effect, but a car mirror is wedge-shaped so the reflections from the front and back surfaces come back to you at different angles. By tilting the mirror at night, you're bringing into play the less efficient front reflecting surface and that's what makes the lights of the car behind appear to dim.

While **travelling** by train,
I've **noticed** something odd.
Objects nearby seem to be **whizzing**
past in the **opposite** direction,
but **objects** further away from the
window seem to be
moving in the same
direction as the train.
How does this happen?

Essentially it's all to do with reference points. It's easy to explain why things near to us are moving backwards – because they are! In comparison to the objects in the background, they're moving backwards very rapidly.

With things further away – usually on the horizon, or at the limit to which you can see – the objects are clearly moving backwards because the train is moving forwards, but we don't see it like that. Our brain rarely sees things as they really are, and tends to rely on comparisons rather than reality!

If I look out of the office window I can see a house about 50m away. In front of the house is a tree. I can't see anything beyond the house. As I walk past the window, the tree clearly moves backwards but the house seems to come with me. If I stick my arm out and point at one of the house windows, though, I can see that my arm does start to move so that it is pointing slightly behind me as I move forward. By pointing at the object, I have given myself a reference point that helps me see that the window is moving backwards. I can't see this normally because there aren't any reference points to compare the house to. However, if the house were transparent so there was something else visible behind it to compare it to, then I would see the house moving backwards just as the tree does.

You can't do this with a train because even if you stick your arm out, the horizon is too far away for you to be able to point accurately.

Why does **grass** seem to be a **lighter** green the further away it is? If you paint a **landscape** **you're** told to paint the background **lighter** than the foreground because that's how it **appears** to the **eye.**

There are lots of atmospheric effects near ground level, and these affect how we see things at a distance. The amount of dust in the atmosphere between you and the object increases with distance, and heat rising from the ground can change the refractive index of the air. Both of these will tend to scatter and smear out the light that you receive from an object. The further the object is away, the more the smudging.

Light from the sun is made up of different colours. Grass close to you will reflect green light and absorb the red and blue light, leaving you to observe the grass as green. The more distant grass will also reflect the same amount of green light, but dust in the atmosphere will reflect white light (all colours) towards you too. This scattering of light basically dilutes the green you see from distant grass.

This is most obvious in cities. If you look out of a tall building, distant buildings will seem paler than closer buildings. They will not, however, seem darker because lots of light is being reflected towards you, just not all of it is a specific colour.

6

Body
Works

Curly Hair, Belly Buttons and Hangovers

What's a **human body** worth?
I mean,
how much would it **cost**
if you **broke** it down
into its individual
elements?

Let's start with the composition of the human body by weight (ignoring some of the smaller trace elements):

	%
	%
Oxygen	65
Carbon	18
Hydrogen	10
Nitrogen	3
Calcium	1.5
Phosphorus	1
Potassium	0.35
Sulphur	0.25
Sodium	0.15
Chlorine	0.15
Magnesium	0.05
Iron	0.0004
Iodine	0.00004

Now, let's assume we have someone weighing 70kg (that's just over 11 stone). That means by weight we have the following amounts of elements:

Oxygen	45.5kg	Sulphur	0.175kg
Carbon	12.6kg	Sodium	0.105kg
Hydrogen	7kg	Chlorine	0.105kg
Nitrogen	2.1kg	Magnesium	0.035kg
Calcium	1.05kg	Iron	0.00028kg
Phosphorus	0.7kg	Iodine	0.000028kg
Potassium	0.245kg		

Now, we need to know the prices for these commodities. We've taken these from a chemical catalogue, and assumed chemicals of average quality because most of us are average people!

Oxygen	45.5kg	£13.66 for 3,264kg	£0.19
Carbon	12.6kg	£6.90 per kilo	£86.94
Hydrogen	7kg	£28.05 for 115.6kg	£1.70
Nitrogen	2.1kg	£15.53 for 2525.6kg	£0.01
Calcium	1.05kg	£3.70 per 25g	£155.40
Phosphorus	0.7kg	£6.90 per 100g	£48.30
Potassium	0.245kg		£339.14
Sulphur	0.175kg		£1.15
Sodium	0.105kg	£17.35 per 100g	£18.22
Chlorine	0.105kg	£68.18 for 33kg	£0.22
Magnesium	0.035		£0.83
Iron	0.00028kg	£4.65 per kilo	£0.001302
Iodine	0.000028kg	£6.00 per 100g	£0.0017

This gives the value of the human body as £652.10.

What is the **strongest muscle** in the body?

The tongue! It is also the only muscle in the body which is only attached at one end.

As far as the other muscles are concerned, the longest muscle is the sartorius, which runs from the hip to the knee, and the largest in surface area is the latissimus dorsi, the broad muscle which covers the back.

I've heard of **people** who can do a **proper** karate 'chop' and split a **brick** in half with a **blow** from the side of the hand. It takes my **builder** a hammer and steel chisel to **achieve** the same **effect.**

Karate is one of the martial arts and requires the body to deliver maximum striking potential with minimum injury. This is achieved through rigorous mental as well as physical training, so don't try it unless you've been taught how to do it properly. There is also some physics involved because speed is the key to this 'trick', remembering that the energy delivered is proportional to the mass x the square of the velocity. Roughly speaking, a properly trained hand will hit a brick at 24mph delivering a force of about 670 pounds. That's not enough to crack a brick if spread over a large area, but because the force is delivered over an area as small as a fist, the brick will crack. Also, the way in which the brick is supported – usually at each end – allows it to yield more easily.

But **bricks** are harder than bones. People break their **bones** all the time but you **never** see bricks crumbling **apart.**

There's been some research done into the real strength of bones which shows they can stand forty times more force than concrete. Hands and feet can withstand even more than that because their skin, muscles, ligaments, tendons and cartilage absorb a great deal of impact. As a result a well-kicked foot can absorb about two thousand times as much force as concrete before breaking.

How can we walk without **thinking about it?** And how much **brain** power does it take?

The simplest questions are always the hardest ones to answer! Walking involves a built-in programme within our central nervous system, and that programme becomes constantly 'modified' on the basis of information from sensory inputs. The basic walking programme probably does not need any 'thoughts', but we constantly make certain changes in our movements owing to external (environmental) and internal inputs which are derived from our intentions.

I am not sure what you mean by brain power. But if you're talking about the number of nerve cells involved, in the case of spiders it has been calculated that of the 30,000 neurones in their central nervous system fewer than 1,000 are responsible for movement. Of course, these form elaborate networks with each other, so it is not a simple affair. Our own nervous system has millions or billions of nerve cells, and it is impossible to say how many are involved in locomotion.

Perhaps more important than any hypothetical calculations is the fact that 'lower' animals can move just as well or as fast as we do even without a large brain. Crocodiles have small brains but tremendous agility, while a house fly has a minute brain but is fully equipped for complex flight manoeuvres, so it's not simply to do with the size of the nervous system or a fly would never get off the ground.

Why do **deep sea** divers speak with **funny** voices?

For divers pure oxygen is a poisonous gas. Even the oxygen in the air – which is only 20 per cent – remains poisonous when 'diluted' by the nitrogen and other gases in the air. However, like all of us, divers need oxygen in order to survive, so they take along tanks of compressed air.

As the diver goes deeper, the pressure on his body increases because the weight of the water above his head increases. The pressure of the air inside his body must also increase, otherwise he would be squashed flat.

The problem is that at high pressure the nitrogen in the compressed air starts to dissolve in the blood, as well as oxygen, and when the diver comes up to the surface again, and the pressure on his body is relaxed, this dissolved nitrogen comes out as bubbles of gas. If the diver comes up slowly, these bubbles appear in the lungs and cause no problems, but if the diver comes up too quickly the bubbles form in the blood vessels and block them, producing intense pain or even death. This is called 'the bends'.

Deep sea divers, who work under huge pressures, avoid this by having something other than nitrogen in their air – they use helium which is very unreactive and does not dissolve in the blood.

Why the funny voice? The speed of sound in helium is about three times that of air and this causes the frequency of the sound produced by the voice to change. The male diver's voice sounds much higher, rather like Donald Duck, and a high-pitched woman's voice is almost impossible to understand.

Why does the helium makes the frequency of the sound go up? It is to do with the way in which humans make sounds in the first place. We force air past our vocal cords causing them to vibrate. We 'choose' the length and tension of the vocal cords so that they resonate at the frequency we want. If the speed of sound in the air we breathe is increased then this resonant frequency is increased, and the pitch of your voice rises. If you ever find a person at a fair blowing up balloons with helium, ask him for a breath of the helium and then try speaking to your friends!

Why are **women's** voices higher than **men's?**

Simply because women and children have shorter vocal cords, and the pitch of a voice depends on the frequency of vibration of those cords, which in turn depends on the tension and length of the cords. So shorter cords means higher pitch.

What causes **hiccups** and are the '**cures**' any good?

Hiccups are caused by sudden contractions of the diaphragm, the main muscle responsible for breathing, which lies somewhere at the bottom of your chest just above your stomach.

When we breathe in, we don't expand the lungs directly: the lungs expand because we increase the volume of the chest cavity. However, the lungs are 'stuck' to the inside of the chest wall, so they have to expand with the rib cage. The diaphragm is the muscle which drive this expansion. When you hiccup, your diaphragm suddenly twitches, and this forces air into the lungs. At the same time, the glottis – a flappy bit at the top of your throat which is part of your voice box – suddenly closes. This sudden closing blocks off the air and causes the noise.

Hiccups don't start in the muscle of the diaphragm itself, but in the nerve that supplies the diaphragm, the phrenic nerve. It's not uncommon for hiccups to start during eating, because the nerves supplying the stomach have links with the nerves of breathing, but hiccups can come on at any time.

There are all sorts of suggested cures, such as standing on your head and drinking from a glass of water, holding up your arms and holding your breath, putting your fingers in your ears and drinking, getting someone to give you a sudden shock, and there are probably many others. It is difficult to know if these 'cures' work, but by the time you've done all the complicated bits and bobs you might have stopped hiccupping anyway!

How much **energy** is consumed by an **active** brain?

We can work this out. The brain uses about 20 per cent of the resting energy of a body. So, if a resting 65kg man's body uses 1.25kcal/min, and a resting 55kg woman uses 0.9kcal/min, the man's brain uses about 0.25kcal/min and the woman's about 0.18kcal/min. In other words, if you're thinking of losing weight by burning more energy, there's no point in just thinking hard about it.

How long would it take a **single** red blood cell to **pass** round my body?

We need some basic information first. Let's assume a body-weight of 70kg. Since 7 per cent of bodyweight is blood, we'll have about 4.9 litres of blood inside us. Each beat of the heart pumps 0.1 litres of blood, and we'll assume you are relaxed with a heartrate of 67 beats per minute. So in one minute your heart pumps 67 x 0.1 litres = 6.7 litres of blood. This means that 4.9 litres of blood gets circulated every 44 seconds, and an average red blood cell takes this long to go round. So the answer is 44 seconds.

If your eyes are **brown** and your **brother's** blue, does that mean you're not **really** related?

Basic eye colour is principally controlled by one gene, either for blue or brown, with brown being the dominant gene. That's why there are more brown-eyed people than any other colour.

We inherit one gene for eye colour from each of our parents. If your brother has blue eyes he has inherited two copies of the blue form of the gene (one from each parent), whereas you have inherited at least one copy of the brown form of the gene. Your other one might be blue, but the brown form of the gene is dominant and wins in the end, making your eyes brown.

Is it **true** that cold **weather** makes us **urinate** more?

Not directly, but when you go out into the cold your body tries to conserve heat by shunting the blood supply away

from the extremities, like the fingers and the toes, and into the centre of the body. One effect of this is to increase the blood pressure in the core of the body – which includes the kidneys – and this leads to higher urine production. Also cold weather does not lead to the same amount of sweating as warmer weather, so the water load has to be excreted somehow. That could make you pee more, too.

Is it **true** that **drinking** too much water **kills** you?

Water intoxication – or getting 'drunk' on too much water – is very uncommon in adults. When it does occur, the symptoms are headache, nausea and lack of coordination. There may also be lapses of consciousness, bloating, abnormally low body temperature and seizures. They are all due to changes in the osmotic pressure in all the tissues as water flows from the fluid surrounding the cells into the cells themselves. This produces two serious effects: the increase in body fluids causes an increase in the intracranial pressure on the brain, which can cause epileptic seizures and even death, and the volume of the blood drops, which may lead to circulatory shock. Several of these symptoms taken together could easily be enough to cause death.

Why do **doctors**
on TV
tap the **knee** of a patient
to **see whether**
their leg jerks?

It's not just doctors on television — all doctors do it. It's to see if the knee jerk reflex is working properly. If it is, it tells the doctor that the nervous system is working. In fact, deep tendon responses like this provide valuable information about the health of the nervous system as a whole.

A reflex is a rapid, automatic response to a stimulus, and a simple reflex involves communication between neurones in the peripheral nervous system and the spinal cord. The brain

may be aware of this passage of information, but does not take part in the actual response. This is what makes the reflex test a good measure of the nervous system on its own. A tap on the knee's patellar tendon with a small percussion hammer causes the thigh muscle that extends the leg at the knee joint to stretch. Receptors, called spindles, within this muscle respond to the change in muscle length and generate nerve impulses. These pass along sensory neurones which carry the signal towards the spinal cord. Here they make synapses (connection points where electrical signals move from one nerve cell to another) and a message is sent straight back down the leg to the thigh muscles. These contract, causing the lower leg to swing forward, giving the familiar knee jerk. But if your nervous system is in poor shape, it doesn't work so well. That's what the doctor is looking for.

What makes farts smell?

The technical name for a fart is 'flatus', and flatus is produced as a result of bacterial activity in the large intestine. These bacteria ferment undigested food, releasing nitrogen, carbon dioxide, hydrogen, methane and hydrogen sulphide. The last three are produced in quite small amounts, but hydrogen sulphide gas is famous for its smell of bad eggs, even in small quantities. This is where the smell of farts comes from. Incidentally, the methane and hydrogen in flatus makes it quite inflammable, and so the stories you may have heard of wild parties in which someone, who was probably very drunk, is persuaded to try to set fire to their farts is not as much of a tall tale as you might have thought. However, what might seem a good idea at the time can prove to be less of a joke later on: the injuries that result are very painful and often need hospital treatment. You then have to explain to the nurse how you did it, which is almost as painful.

The composition of fart gas is highly variable. Most of the air we swallow, especially the oxygen, is absorbed by the body before the gas gets into the intestines so what reaches the large intestine is mostly nitrogen. Bacterial action also produces hydrogen and methane. But the relative proportions of these gases that emerge from our anal opening depend on several factors: what we ate, how much air we swallowed, what kinds of bacteria we have in our intestines, and how long we held in the fart.

The longer a fart is held in, the larger the proportion of inert nitrogen it contains, because the other gases tend to be absorbed into the bloodstream through the walls of the intestine. So a nervous person who swallows a lot of air and moves stuff through his digestive system rapidly may have a lot of oxygen in his farts, because his body doesn't have time to absorb the oxygen.

Why do we get cramp, and why does it **always** seem to be in the **feet** and lower **legs?**

Cramp is a sustained and painful abnormal contraction of a muscle or group of muscles, and is a form of hypertonia or excessive muscle tone. It is caused by greater than normal levels of alpha neurone activity, and this keeps the muscles contracted despite your best efforts to relax them. Basically, this means the nerves controlling muscle contraction are continually sending the signal for the muscles to contract, even while you are willing them to relax, and they are winning.

Sometimes you get cramp after exercise. This is because there is not enough oxygen getting to the muscle, so they respire or 'breathe' anaerobically, which means without oxygen. They do this to gain enough energy in order to contract. But without oxygen the 'breathing' process causes a build-up of lactic acid, which causes the ache, and without oxygen this lactic acid cannot be broken down.

If you are fit, the number of muscle fibres in your muscles increases, as does the blood supply, and this allows more oxygen to the muscles so they can exercise for longer before becoming tired. It can also remove more lactic acid so there is less chance of cramp. If you are so fit that you never get to the stage when your muscles can't get rid of the lactic acid that builds up when you exercise, you won't get cramp.

We get more cramp in our feet and legs because our circulation is less efficient in our legs compared with the rest of our body. Because of the decreased blood supply, less oxygen reaches our legs, making cramp more likely due to the build-up of lactic acid.

Why do we **shut** our eyes every **time** we sneeze?

Because it's physically impossible to keep your eyes open when you sneeze. Sneezing is a reflex reaction controlled by the autonomous nervous system, which controls your heart rate and breathing, and can't be consciously controlled. There used to be a theory that if your eyes didn't close when you sneezed then your eyes would pop out, but nobody had been able to test it. But that's the only explanation, because scientists have come up with no other.

Why are our **fingers** **different** lengths?

When fingers first form in the human embryo they are all about the same size but each one has a special 'code' or identity. At this point each finger is about 1mm long, and made of cartilage cells which are programmed to grow.

Because each finger has its own special identity, evolution can programme the fingers to grow independently by employing a special 'signalling molecule'.

Each finger is exposed to a different 'concentration' of the signal, and that makes the finger longer or shorter than the others. The thumb is least under the influence of this signalling molecule, and ends up shorter.

That's the 'how', but the 'why' is more difficult to answer. Perhaps it gives us a greater variety of fingertip grip possibilities; possibly it is because when most people curl their hands, the finger tips come to about the same relative position – try it.

Having said all that, human fingers are pretty uniform compared to some species. Bats have fingers which are enormously long compared to the others; pterodactyls are the most spectacular with one huge finger and three tiny ones.

If you **cut** your finger at the **tip,** and the **fingerprint** is damaged, would it grow **back into** the same fingerprint?

Fingerprints are produced by 'friction ridges' which are there to help us grasp and hold on to objects. If these 'ridges' are damaged by a cut, the depth of the injury determines whether or not you get your entire fingerprint back. A deep cut would result in a scar which, of course, is not the same as your original fingerprint. But if the cut is a shallow one, then the ridges and furrows would grow back in their original pattern – it never changes, no matter how deep the damage.

Why does my **tummy** sometimes **rumble?** It always **seems** so loud, can other **people** hear it?

Your tummy doesn't rumble *sometimes*, it rumbles *all* the time, and not just when it's empty. Borborygmi – the technical name for stomach rumbles – is caused by the movements of gas. We swallow air when we swallow food, and as our stomachs contract the air is pushed around. The stomach tends to make more rumbles, or stronger contractions, when you are nervous or hungry, and these tend to be louder. But don't worry – you are far closer to your stomach than anyone else and the sound is transmitted to your ears via bones and muscles. They would have to be rumbles of earthquake proportions to become really distracting to anyone else.

What **makes** your hair curl?

It's a simple question, but science still has no real answer. As usual, however, there are various theories.

We know what influences the curliness or straightness of hair: genes, metabolism (body chemistry), racial influences, diet, illness and possibly stress or shock. Also, what happened in the womb can determine it.

It was thought that the curliness of the hair was due to the shape of the follicle: a straight hair came from a straight follicle, while a curly hair came from a curved follicle, but this couldn't explain how an individual's hair could change from being curly to straight or vice versa.

Hair growth depends on cell division in the papilla which is found at the base of the hair follicle. If you think of the growing hair as a clock face, and if the cells dividing at each hour divide at an equal rate, the hair will grow straight upwards. If the cells at 3 o'clock are dividing faster than the rest, the hair will bend towards 9 o'clock as it grows. If the cells then began to grow faster at 9 o'clock the hair would bend back towards 3 o'clock, and you would get wavy hair.

Tight curls are formed when the hair cells divide faster in a cycle 'around the clock'. If the cells in the follicle of someone with curly hair suddenly started to divide at an equal rate, the hair produced would become straight.

Do **identical** twins have **identical** fingerprints?

No. Even monozygotic twins (identical twins from the same egg) have slightly different fingerprints.

Fingerprints form before birth and their shape is thought to be influenced by both nutrition and the growth of fingers during the thirteenth week of pregnancy. As the fingers form, pads of skin develop on the ends of the fingers which eventually develop ridges. Foetuses with higher blood pressure will have swollen finger pads, so the patterns formed are more likely to be whorls. Although fingers will become scarred or blemished during life, the patterns themselves remain unaltered. Fingerprints are always unique, not only to the individual but to the digit as well. Certain matching patterns do often exist in the prints of twins, but there the likeness ends.

Is it **possible** for one identical **twin** to be **left-handed** and the other to be **right-handed?**

Some argue that handedness is a genetic trait. According to the figures, if both parents are right-handed, the probability of their child being left-handed is just 9.5 per cent. If one parent is left-handed, this figure jumps to 19.5 per cent and if both parents are left-handed it reaches 26.1 per cent. So if it is true that handedness is genetically inherited, identical twins would always show the same handedness as they share exactly the same genes (or genotype).

But some argue that a child is trained to be left- or right-handed, or possibly that handedness is influenced by conditions in the womb such as unusually high levels of testosterone. If this were true, it is possible that identical twins could develop different preferences for handedness.

On the present evidence, it would appear that identical twins have the same handedness.

How
do your **fingernails** grow?

If you look at your fingernail, you will see that the bottom of it is buried in your finger. This part of the nail is called the nail bed, and this is where the nail does its growing. Cells divide in the nail bed to form new nail cells, and they push the older cells towards the tip of the finger. Once the nail emerges from the nail bed, the cells have died and been covered with keratin which is a very tough protein which helps to protect our fingertips from damage. This is how the nail grows, by new cells at the bottom pushing the cells at the top towards the end of the finger.

Almost all body cells are formed by a process called 'mitosis'. Mitosis is when a cell copies itself to make another, identical cell. First, the genetic material is copied, and then the cell makes extra of everything else. Then the cell tears itself in two, and each part becomes a whole new cell. This is how the body grows and and renews itself, including your fingernails.

How **long**
do your **nails** grow in a month?

Fingernails grow at a rate of 0.5mm each week. Since there are (52 divided by 12) 4.33 weeks in each month, your fingernails will grow 2.16mm each month. They will grow slightly faster during the summer months and slightly more slowly during the winter months. Your toenails grow at a marginally slower rate than your fingernails.

What makes urine **yellow?**

Urine is part of the body's fiendishly clever waste-disposal system, managed by the kidneys whose job is to keep the salt content of the blood constant and to filter waste from the bloodstream. So your urine consists of water, salt water and waste products the body wants to get rid of.

The major waste product is ammonia, which comes from the body cells, while from the blood comes bilirubin which is created when haemoglobin breaks down. These substances are hazardous to the body, so the kidney converts the ammonia to urea and degrades the bilirubin to urobilogens, which are yellow. These give the urine its colour. But if you drink sufficient water you can dilute the urobilogens. This is why the colour of urine in a dehydrated person is very yellow.

I know that we shed skin every day, but how much?

True, we do shed skin every day and make a real mess as a result. Every minute we shed 30,000 to 40,000 microscopic skin cells, which adds up to a staggering 4kg of dead skin that we lose every year. Some of it drops off of its own accord, but much is lost by rubbing against things, even our clothes. And where does the dead skin end up? You need look no further than house dust.

But don't worry. New cells are continually being formed to replace the ones that fall off. The top layer of your skin, the part of it which you can see, is called the epidermis and is made up of four or five distinct layers of cells. The palms of your hands and the soles of your feet are normally exposed to greater friction than the rest of your body so they have an extra epidermis cell layer.

The dead skin cells fall off the top layer of your epidermis, the *stratum corneum*, which is made up of twenty-five to thirty layers of flat and tough dead skin cells. The bottom layer of the epidermis, called the *stratum basale*, contains cells which continually divide, producing new cells which work their way up through all the layers of the epidermis, a bit like people moving to the front of a queue.

A skin cell has a short life: about two to four weeks after the cells have been formed, they die and await collection by the vacuum cleaner.

If we are **constantly** shedding our skin, then why do **tattoos** remain?

The human skin has two layers: the outer epidermis and inner dermis. The outer layer is about four or five cells thick, whereas the dermis is much thicker. To create a tattoo, dye is injected deep into the dermis cells in the lower level of the skin. The dermis is relatively stable and changes very little over a lifetime. In the outer layers the cells are completely replaced, but down here individual molecules will be replaced but not entire cells. Once you've got your tattoo, you're stuck with it and body chemistry is not going to help you lose it.

Do **bald** people get dandruff?

Yes, they do. Bad luck, isn't it? Dandruff is caused by bacteria, yeast and fungi on the scalp, and these can act on the scalp with or without hair being there. However, dandruff is more common in people with hair because the thatch helps trap both heat and water which provide ideal living conditions for bugs and suchlike.

155

Why do my hands smell after holding coins?

There's a lot of chemistry happening here, and most of it is to do with the reaction between the sweat on your hands and the metals in the coins.

The composition of your sweat varies with the food you eat. If you are eating a protein-rich diet then your sweat will be high in compounds containing nitrogen, such as ammonia, and these form new compounds when in contact principally with the copper which coins contain.

Not all coins will produce the same smell in all people. If you were to put a coin into the hand of a fit athlete (who we assume has been gobbling up protein to provide strength and stamina) then their hands would smell stronger than those of an idle person on a meat/cheese-free diet. Also, a male athlete might have a stronger reaction with the coins than a female because the higher level of the hormone testosterone decreases his body's level of acidity, as shown by the increased nitrogen compounds in the sweat.

Does your nose run in space?

According to NASA, astronauts frequently complain about having stuffy heads, particularly during their first few days in microgravity. It may be because the fluid in the astronaut's legs and abdomen moves up to the chest and head in a weightless environment. However, a runny nose cannot happen in space because there is no gravity to pull the fluid down. Instead any excess fluid stays in the astronaut's sinus cavities until they blow their nose, then the pressure forces the fluid out.

Is eating **snot bad** for you?

I shouldn't think so. In fact we are eating it all the time. Mucus — the proper name for snot — is produced by cells lining the respiratory tract and we are constantly swallowing the stuff as it's slowly moved along to the back of the throat by tiny hairs called cilia. The mucus, far from being a pest, is a defence mechanism designed to trap pollen, dust and bacteria that are present in the air we breathe, and it is better that it ends up in the stomach than your lungs.

I suppose it's just possible that eating it could be harmful if it had trapped really noxious particles from the air, but that risk is small and the conditions in the stomach are pretty tough and make the survival of harmful germs unlikely. So, given that the vast majority of the mucus you produce is swallowed anyway, taking the short route to your stomach doesn't make a difference in the end.

Is there any **reason** why most **opera** singers are so **large?**

There is a theory that being overweight could benefit the voice. Numerous parts of our body interact to produce the sound we know as our voice, but the most important is the larynx or voice box.

Our voice comes from vibrating vocal folds in the larynx which have an outer surface called the mucosa which cushions the collisions between the vocal folds as they vibrate. Some research has suggested that with a thicker and fatter mucosa the voice is more efficient at converting airflow from the lungs into a louder, more powerful voice. If being overweight makes you deposit more fatty tissue in your mucosa, then this might make your voice more powerful.

Why do some **people** have belly buttons that **stick out,** and **some** people don't?

It is all determined in the few weeks after you are born. The umbilical cord, which connects mother and baby in the womb, supplies oxygen and nutrients. This cord is cut shortly after birth, and it is the way in which the hole through the middle of the cord closes which decides if the eventual belly button sticks in or out.

When the abdominal muscles do not close up completely, you end up with an 'outie' belly button. If it does close completely, then you will have an 'innie'. The way the umbilical cord is cut after birth can also play a role in the future shape of a belly button. You are more likely to get an 'innie' if there is a good cut and only a small section of the cord left. It will form an 'outie' if a longer section of cord is left.

Why does **alcohol make** you feel drunk? And then why does it make you **sick?**

Alcohol, which is a poison if taken in large quantities, gets to work on brain cells called neurons, and affects the way the brain uses three particular chemicals: gamma-aminobutyric acid (GABA), serotonin and dopamine. These three are all neurotransmitters, which means that they pass between different nerve cells as signals, activating or deactivating the cells for which they are targeted.

Alcohol tends to lead to an increase in the levels of serotonin. This causes a happy sensation, and is one reason why drinking is immediately pleasurable to the drinker. GABA, on the other hand, generally inhibits and slows the brain down, which contributes to the feeling of drunkenness. Dopamine is another pleasure-regulating chemical, but is also responsible for coordinated movement, which may be why you start to stagger when the alcohol takes effect – and why you must not drive.

In large quantities alcohol can damage various organs, including the liver, and the body is essentially aware of its toxic effects. It will tolerate a small amount, but when large quantities are consumed the reaction is to vomit. Of course, this may not remove much alcohol from the body – it's possible that by this point most of it has already been taken in. So then the only option is to recover in the usual way, with a hangover.

Why do we **always** crave starchy and **high-fat** foods when we are **hungover?**

Alcohol does several things to your body, and all of them end up producing a feeling of hunger. First of all, it mimics the effect of insulin and reduces your blood sugar level. This is the classic way of signalling to the body that it is hungry, and you act accordingly.

Alcohol also stimulates the flow of saliva and gastric digestive juices – the so-called 'aperitif effect' – which some scientists think may also add to the feeling of hunger.

Alcohol is also a diuretic, which means it stimulates the removal of fluid from the body leaving you short of water or dehydrated. If you've drunk enough to have a hangover, you are probably suffering from severe dehydration so the hypothalamus creates a sensation of thirst. The sensation of thirst and hunger are often confused because they are both caused by stimulation of the lateral hypothalamus, that portion of the brain which controls body temperature, thirst, hunger, water balance, emotional activity and sleep.

To get rid of the hangover feeling we crave food, and there is no better way of feeding the body quickly than by stuffing it with fats. Fats dissolve quickly in the mouth, releasing their flavours, but they also have a way of making the flavours last longer so you can hold on to the taste and gain satisfaction long after the food has left your mouth. Also, it is thought that foods high in fat and sugar stimulate the production of endorphins, which are the body's natural painkillers and cause a pleasurable sensation when they are released. Perhaps good for bad heads, too!

Why do the **bubbles** in **champagne** make you get drunk more **quickly?**

Alcohol is quite a small molecule and is absorbed quickly into your bloodstream. The bubbles, which are of carbon dioxide, allow it to be absorbed even more quickly by stirring the alcohol around your mouth, stomach and intestines. In an experiment where people drank flat champagne and fizzy

champagne, the flat drinkers ended up with almost half the amount of alcohol in their bloodstream.

You can reduce the effect by drinking champagne from a broad, shallow glass. A tall, narrow flute allows less carbon dioxide to escape and maintains the potency.

How does your **body** gain more weight than the weight of the **food** you actually eat? If you ate 1kg of **chocolate**, would you **gain** more weight than the 1kg of chocolate **eaten?** **If you ate** 1kg of apples, would you gain **less?**

You can't put on more weight than the weight of the food you eat. This would violate the laws of thermodynamics and the conservation of mass and energy. Also you use some of the energy contained within the food to digest and process it within the body.

It's very difficult to calculate how much weight you would put on from eating 1kg of a certain type of food. First it depends on your metabolism, and individuals vary in the way their metabolism works and the rate at which they use food. Metabolism is the balance between the amount of food broken down and used for energy and protein synthesis, and the amount of food which is stored around the body. The balance between the two is affected by factors such as your bodyweight, the amount of energy used to exercise or keep warm, and age – metabolism is slower in older people.

So, someone might gain absolutely no weight from eating 1kg of chocolate while others will gain some weight. We can't say how much weight someone would gain from eating that much chocolate as everyone expends different amounts of energy in a day, but we do know how much energy 1kg of chocolate contains. This is how it is worked out.

The four major components of most foods are carbohydrate, protein, fat and water. Foods also contain vitamins and minerals, but in much smaller quantities. The energy content of different foods will depend on the relative amounts of carbohydrate, protein, fat and water found in the food.

On the back on food packets you will see the energy content is given either in calories or kilocalories (Kcal), which are used interchangeably even though a 'Kcal' simply means 1,000 calories. One calorie is the amount of energy required to raise the temperature of 1ml of water by 1°C, at 15°C. When people talk about the number of calories in food, however, they are actually talking about kilocalories, which (perhaps confusingly) are generally called calories for short.

An average 100g milk chocolate bar contains about 7g protein, 54g carbohydrate, 34g fat and 5g water and will give you 550Kcal of energy. 100g of apples will contain on average 0.2g protein, 15.4g carbohydrate, 0.35g fat and 84g water and will give you 60Kcal of energy.

The average adult male needs 2,500Kcal per day and if they managed to eat 1kg of chocolate they would be consuming 3000 extra calories which the body would store as either fat or carbohydrate reserves.

How long
can a human stay
awake?

The official world record for staying awake is 264 continuous hours (eleven days) and was set by Randy Gardner, a 17-year-old student, in 1964. He was monitored by sleep specialists throughout and apparently suffered few or no negative consequences. Other research subjects are known to have remained awake in carefully monitored laboratory settings for eight to ten days.

Although none of these individuals experienced serious medical, neurological or physiological problems, they all showed reduced concentration, motivation and perception as sleep deprivation increased. Brief episodes of altered consciousness (known as microsleep) became more frequent, leading to a loss of cognitive and motor functions. This means that although we can stay 'awake' for several days we end up in a cognitively dysfunctional state.

Can you keep someone
awake long enough
to kill them?

Yes, you can! An experiment was conducted using rats on an enclosed turntable that began spinning whenever the rodent's brain waves suggested it was beginning to nod off, forcing the rodent to stay awake. After about a week of this, the rat starts showing some signs of strain: lesions break out on its tail and paws, it becomes irritable and its body temperature drops as it attempts to make itself warmer than usual. It eats twice as much food as normal, but loses 10 to 15 per cent of its body-weight. After about seventeen days of sleeplessness, the rat dies. This suggests that sleep is nearly as vital to life as is food. It is very likely that something similar would happen in humans.

Why are some people's lips red, and others have more of a pink colour to them?

The surface skin on our lips is more translucent than the skin on our faces because it contains less keratin – a tough protein substance which is a major part of skin, nails and hair. This allows the tiny blood vessels that lie under our lips to become more visible, and this accounts for the red/pink colour. The intensity of the colour will depend on the individual skin thickness and the quantity of blood vessels that run under the lips. The more blood vessels and the thinner the skin, the redder the lips.

Lip colour is also affected by the presence of melanin – the pigment which gives skin its colour. Although the amount of melanin in our lips is much less than in the rest of our skin, the more melanin people have the more this pigment tends to cause the lips to take on a purple to brown colour. The amount of melanin in our skin is inherited, and genetics will play a role in determining lip colour. However, you need to bear in mind that there are probably many genes influencing skin colour and its thickness, so you can't guess what a child's lips will look like just by looking at the parents'.

Why do we **blink?**

There is more to blinking than meets the eye. Clearly we have to blink to cleanse and moisten the eye: each time the eyelids close, salty secretions from the tear glands are swept over the surface of the eye, flushing away small dust particles and lubricating the exposed portion of the eyeball. Normally we blink every four to six seconds, but in irritating conditions such as a smoke-filled room we blink more frequently to keep the eyes clean and moist.

However, we blink more often than we need to, if blinking were just about keeping the corneas of our eyes moist and clear. Infants only blink once every minute or so, but adults blink an average of ten to fifteen times a minute. Scientists now think this is all to do with gathering information, because experiments have shown that we blink less when information is coming at us thick and fast, and we blink more often when we are not taking much in.

Blinks are like punctuation marks of the mind, signalling a pause in the activity in your head. If we are reading interesting material, we blink an average of three to eight times per minute, as opposed to fifteen times per minute when we are not engaged in an attention-demanding activity. We are also most likely to blink as our eyes shift from one page of text to the next, or from the end of a line of text to the beginning of the next line.

Any single blink isn't always like the next. Scientists have shown that frequency and duration vary under different conditions. Royal Air Force pilots flying simulators over 'friendly' territory have been shown to blink more frequently and to keep their eyelids closed for longer than when flying over 'enemy' territory. Pilots blink the least when they have been spotted by enemy radar and are attempting to find and evade missiles, or while landing an aircraft.

How much of our **lifetime** do we **spend** with our eyes closed, just by **blinking?**

A blink lasts about 0.3–0.4 seconds. We blink about five times a minute, every minute for about eighteen hours a day. This adds up to half an hour a day, which is about five years in an average lifetime.

Why can **babies** breathe and swallow at the same time, **but adults** can't?

We have two separate tubes in our throat: food goes down the oesophagus into our stomach, and air reaches our lungs through the larynx. At the top, near our mouth, both tubes are connected. The problem is that if food enters the airway and blocks it, we can choke to death. This is why we have developed a reflex that does not allow us to breathe and swallow at the same time.

Babies under the age of six months don't have this reflex and therefore have the ability to swallow and breathe at the same time. So why doesn't that make it dangerous for babies? The answer is that in young babies the larynx is much higher in the throat than in adults, and this means that when they are suckling, the milk can run around each side of the larynx into the oesophagus without getting into the lungs. As the baby grows and the shape of the larynx changes, the reflex develops. No one really knows how this happens, but it seems that not being able to breathe and swallow at the same time is the 'normal' state which is 'switched off' during the suckling period.

If there's **so much** water in our **bodies,** why do we appear **mostly** solid?

The adult body contains approximately 55 to 60 per cent water, but some parts of the body are richer in water than others. The brain and the skin have 70 per cent water, blood is 82 per cent water and the lungs are nearly 90 per cent water.

We appear mostly solid because the water will be contained either inside our cells or inside organs, and without it the chemical reactions necessary to keep us alive would not be able to take place. The fluid also gives shape to the cells which means we would appear rather shrunken if all our water was removed by freeze-drying.

There is also a lot of water in our blood mixed in with a variety of blood cells such as haemoglobin, white cells, platelets, etc. Water is what makes blood a fluid and enables it to travel to all parts of our body through the blood vessels, and carry out its vital biological functions.

If you **drink** some water while **standing** on your head will it go to your **stomach?**

Anything that you eat or drink will end up in your stomach, regardless of the position that you are in. Food is not pulled down to the stomach by gravity, but through a series of reflexes controlled by the brain.

The mouth leads not only to the stomach but also to the nose and lungs, so it is important that once we swallow food or drinks these don't end up in the wrong place. Swallowing triggers a reflex where the only pathway that remains open is the oesophagus, the tube that connects your mouth with your stomach. Muscles in the oesophagus start contracting to make sure that the food and drink go in the right direction towards the stomach and this will happen even if we are standing on our heads. Occasionally, the reflex fails if we eat and talk at the same time, and a tiny bit of food or drink will go into the wrong place making us choke.

This swallowing reflex is also why astronauts are able to eat in the absence of gravity. Even if they are floating inside their spacecraft their food will end up inside their stomachs.

Are newborn **boys** more **fragile** than newborn **girls?**

You'd think there wouldn't be much difference, but male newborns *are* more fragile than females.

There are only theories as to why this might be. Some argue that it is possible that the hormonal environment experienced in the womb has more negative effects on the development of males. This is because males, in order to overcome the influences of oestrogen produced by the mother, have to start producing testosterone as soon as possible, which requires a quick development of the testes. In order to achieve this, male foetuses have a higher metabolic rate than females which could make them more vulnerable.

It is also possible that environmental pollutants such as PCBs and detergents may be mimicking the female hormone oestrogen and damaging the male's reproductive system as it develops in the womb. However, nature seems to recognize this as a bit of a problem and boys' vulnerability is compensated for by an increased likelihood of males being conceived, compared to females. On average, 125 males are conceived for every 100 females, and although more male foetuses miscarry, more boys are born – roughly 105 boys to every 100 girls.

It also seems that more boys are conceived at a time of year when conditions for pregnancy and birth are optimal, which would be another way of ensuring that the differences caused by males being more fragile are overcome.

What is **ear wax** for and **why** does it **taste** so bad?!

Have a look at someone's ear and you'll see the external auditory canal, which is the curved tube that runs from the eardrum to the outside of the head. It contains a few hairs and glands which produce cerumen or ear wax. The hairs and wax together help prevent dust and dirt entering the ear.

Normally you produce just enough ear wax so your ears shouldn't need cleaning. In fact cleaning them may encourage the glands to secrete too much wax. Only when there is something wrong with the ear – an infection, for example – do we produce too much wax, and then we may need to consult a doctor to have our ears cleaned. Note: you should never clean *inside* the ear canal, and only use cotton buds to clean around the external bits of the ear.

The wax itself is a mixture of desquamated keratinocytes, otherwise known as bits of dry skin and hair, combined with

the secretions of both the ceruminous and the sebaceous glands of the external ear canal. The major organic components of ear wax are long chain fatty acids, both saturated and unsaturated, alcohols, squalene (a chemical also found in the liver oil of sharks) and cholesterol.

Why does it taste so bad? Ear wax contains long chain fatty acids which are the same molecules found in butter and margarine; when these fatty acids are exposed to oxygen in

the air they become oxidized, which leads the butter/
margarine to become rancid. The same thing happens to ear
wax – it's rancid!

Why do we have **freckles** on the **backs** of our hands but **none** on our **fingers?**

Freckles are clumps of skin cells with more melanin in them
than other cells and tend to form when our skin is exposed
to sunshine. As we walk we tend to hold our hands with our
fingers curled inwards, shading them from the sunlight and so
reducing the chances of freckles forming.

How long can a dead **embalmed** body stay in a house **before it** starts to **decompose?**

Bodies that are mummified, or embalmed, can last decades or
even centuries before decaying. The Egyptians mummified
corpses by drying them, creating an inhospitable environment
for the microbes to live and reproduce in. A salt mixture
called natron, which is found on the banks of the Nile, was
also used to dry out the bodies by making them more
alkaline, which bacteria find unfavourable. However, the dry

North African climate was one of the main reasons that the Egyptians were so successful at preserving their dead.

Formaldehyde, phenol, methanol, ethanol and other solvents are generally used in modern embalming. The blood is forced out of the dead body by injecting embalming fluid into the circulation system using a pump. The embalming fluid normally contains a disinfectant – phenol – that kills the microbes present in the body, and a preservative – formaldehyde – that 'fixes' the cells. All biological activity is stopped when formaldehyde is added as it cross-links proteins and other molecules, freezing the structures into position. This method of embalming can be used to halt decomposition for decades.

What stops you **rotting before** you die?

There are within our immune system white blood cells, antibodies and antioxidants, which are present and active throughout our body while we're alive. They're at work not just in our blood, but between the other cells in certain areas of the body too. Their job is to detect anything 'foreign' and kill it.

As soon as we die, the oxygen supply to all our cells, including the ones in our immune system, stops. This means the microbes in the body can start growing freely. Soon the whole body becomes a 'feeding ground', helped by the death of our other body cells which fail to hold their shape and allow their contents to spill out, creating a 'soup' on which the microbes feed. By this point decomposition of a corpse is well under way.

While we are alive, the skin is also a defence against rotting, acting as a physical barrier to microbes. But once dead, the skin loses its structure and that defence is lost.

The process of decay is actually remarkably fast. In hot, humid conditions decomposition can be under way within a day. In cooler, more sterile conditions (such as a morgue) the process slows down and can take months.

Is it possible to live for ever?

Let's first deal with a theoretical world, and then the real world. According to Einstein's theory of relativity you can't travel at the speed of light, but the closer you get to it the slower your time seems to go compared to people on the Earth who aren't travelling at the speed of light. In theory, you could slow your time down by travelling at this speed until everyone on Earth is dead. However, to you it would still feel as if your life was passing at the same rate as if you had stayed at home, so there would be no personal feeling of having lived for ever.

From a biological point of view, there are many reasons why we don't live for ever. One is that there are cells in our bodies which don't reproduce (for instance, nerve cells, brain cells and bone cells). So when they die through general wear and tear, they're not replaced. Also, when cells which do reproduce split and make copies of themselves, errors can creep in during the duplication which give rise to mutations. Every generation is liable to error. So the longer you live the more copies are made, and more copies means more mutations, until eventually there are not enough 'fully working' cells left to do the job of keeping you alive.

7

Kitchen and Home

Jelly, Diamonds and Custard Powder

I've made **countless** pineapple jellies
for my kids over the **years**
using tinned **pineapple.**
Now, because I thought **fresh**
would be **healthier,** I used a
freshly chopped **pineapple** instead.
All I've ended up with
is a bowl of slop,
and certainly **not jelly.**
What happened?

It's simply that enzymes are out to spoil your entire party. Pineapples contain an enzyme called papain which can cause proteins to break down into small fragments. Gelatin, which is the stuff that makes jelly wobbly, happens to be a protein which papain is only too happy to break apart. The result is that the jelly will never set. What's so different about canned pineapple? It's simply that part of the canning process involves heating the pineapple which destroys the papain. The result is that the gelatin protein remains untouched, and wobbly perfection is the end result.

Don't think that papain is entirely a pain. Its ability to demolish protein also makes it capable of tenderizing meat whose toughness is caused by the connective collagen, which is a protein. Also suspended protein in freshly brewed beer can be cleared with papain.

Be warned — don't make jelly with fresh kiwi fruit, figs or mangoes. It won't set because they contain papain too.

I've got a bunch of **bananas**,
only a **couple** of days old,
which have gone black.
Was I sold **rotten** fruit?

No, probably not. It's enzymes again – they really don't seem to like you very much. For a start bananas are a tropical fruit which have never known anything but the hot glare of the sun, until they met the inside of your fridge. They simply are not built for the cold, unlike apples or pears which will happily sit in a fridge for weeks. In the case of bananas, the membrane which encloses the cells breaks down and leaks, and what emerges are enzymes marching in search of something to destroy. One enzyme, known as polyphenyloxidase, reacts with tannins which are normally confined to a separate part of the cell, and the reaction between the two causes the formation of brown compounds – or black-looking bananas to you.

The ideal temperature for storing bananas is 13.3°C. Below 10°C bananas will start to blacken, so wrap them up well on chilly nights.

Why are **ice cubes** always cloudy, when the **water's clear** to start with?

For three good reasons, all fine examples of the things that can happen when you start putting obstacles in the way of a beam of light. First, your ice cube is not one big crystal but lots of little crystals, and that provides plenty of opportunity for the light to be *diffracted* as it hits the edge of the crystals. Secondly, the atmospheric gases such as carbon dioxide, oxygen and nitrogen are more soluble in cold water and as the water cools towards freezing bubbles of gas become trapped. They might be very small bubbles, but they're still big enough to *refract* the light. Thirdly, small pockets of liquid water can remain unfrozen, even within an ice cube – another *refraction* opportunity. Add all three together and you see there's little opportunity for the light to pass through an ice cube and come out clean the other side.

The difference between *diff*raction and *re*fraction? Diffraction is what you see when a light wave bends round the edge of an obstruction; refraction is the bending of light as it passes from one medium to another.

If you take two cups
of **coffee,**
one at 40°C, one at 30°C,
and **place** both
in a freezer,
which will **freeze** first?

It defies common sense, but it's the hotter one that will freeze first because the hot water molecules have enough energy to leave the body of water as steam, taking heat energy away from the water. Cold water molecules don't have enough energy so don't leave the body of water as often. So although the hot water is hotter, it loses its heat much faster because its molecules have more energy, so the hot water loses energy quicker than cool water, catches up, then overtakes and reaches freezing point first. This has been known about for a long time. Aristotle (384–322 BC) wrote in his *Meteorology* that, 'Many people, when they want to cool water quickly, begin by putting it in the sun. So the inhabitants when they encamp on the ice to fish [they cut a hole in the ice and then fish] pour warm water round their rods that it may freeze the quicker; for they use ice like lead to fix the rods.'

When you hold
a cup of **coffee** with a drop of milk
in the centre and then **turn** round,
the drop of milk **stays**
still and the coffee moves around it.
Why?

Do you ever suffer from inertia? It's that feeling that no matter how much people try to shove you around, or what pressures the world places upon you, you simply aren't going to budge. Well, coffee's got inertia too; it's a tendency to stay put. When you waltz around with the cup of coffee what happens is that the cup is yielding to the force you are applying directly to it, but not the coffee which wants to stay exactly where it is. What causes it to move around just a little is the friction between the coffee and the cup. But because there's no friction between the milk and cup, the drop of milk stays exactly where it is.

I've noticed,
when making **hot drinks**,
that **just before** the kettle comes
to the boil, it goes **quiet.**
Why is this?

As water is heated, dissolved gases in the water start to come out of solution, which is the gentle hissing that first gives the game away that things are hotting up. As water approaches boiling point, all the dissolved gases have been released so there is no more bubbling – this is when the kettle goes quiet. Enjoy the peace while it lasts, for as the water begins to boil, the convection currents in the water become very violent and the water becomes noisy again.

I've **noticed** that if you take a cup of **coffee** straight out of the **microwave** and plunge a spoon into it, it boils **straight** away. Is it **magic?**

Answer: I'd call it dangerous. You must be very careful doing this as people have been seriously scalded. You have to understand a little of the way microwaves heat food, and then you'll see the dangers. Unlike conventional heating, as in a saucepan, microwaves don't penetrate very far into whatever

they are heating. The result is that some parts of your cup of coffee could be well above boiling point while other layers might still be cold. What happens when you plunge your spoon into the cup and stir things up is that the cold layers can suddenly be brought to boiling point, or even higher, by sudden contact with the superheated ones. Instead of escaping slowly as in conventional boiling, the steam that is produced, erupts from inside the coffee and if you happen to be in the way of it, the scalding coffee will end up all over you. It can hurt.

A microwave makes the water **boil** by giving **extra** energy to the water **molecules.** So, **if I were** to give them extra energy in another way, say by **banging** the cup up and **down** on the table, and I did that for **long enough,** would the **water** come to the boil?

Do you seriously think those guys at Panasonic who make the microwaves have been keeping this a secret from us all these years? Microwave radiation is at just the right frequency to cause maximum vibration of the water molecules, which causes the water to heat. In a kettle, it is the transfer of energy from the heating element that gives the molecules of water on the surface sufficient energy to change from liquid to gas and escape as steam. Banging the cup down on the table does indeed transfer energy, but mostly to the cup which *theoretically* will warm up. The same is *theoretically* true of the coffee, but it's going to be minute because the transfer is inefficient. So, if you want a cup of theoretical coffee, bang the cup up and down for a millennium or two. Otherwise put the kettle on.

If you want hot milk in your **coffee**
you've got to **watch** over it
or it **rises up** the pan
and makes a disgusting **mess.**
What **makes** milk such a pest?

First, you've got to understand that milk is packed with good food: it's got vitamins and essential fats for body growth and, most importantly when it comes to boiling the stuff, it's got proteins which are made up of long molecules that contain amino acids – the basic building blocks of body cells. When the milk is heated, these proteins unfold and wrap themselves around the bubbles of air that are trying to rise up in the milk. It stops them escaping as quickly as they would do in water, and you end up with a mass of bubbles which can't get anywhere except up the pan and over the cooker you have just cleaned. So, if milk weren't so nutritious, it wouldn't be such a pain.

But surface tension matters as well. The surface tension of water is very high: think of the surface of water as a skin of tightly stretched rubber that is always eager to return to its original shape after a bubble of air has passed through it — it sort of snaps shut behind it. Milk, on the other hand, has a low surface tension, so the bubbles survive longer than they would in water. Add that to the protein-wrapping ability they already have and you see why the bubbles of air in boiling milk don't give up without a fight.

Incidentally, if you want to prove it's all to do with surface tension, add some washing-up liquid to a pan of water and bring it to the boil. It will behave in the same way as milk.

Why do Rice Krispies go 'snap, crackle **and** pop' as soon as I pour on the **milk?** And why do they **stop?**

It's all to do with trapped air. To prove it, take hold of a Krispie, which is simply made of expanded rice, and break

it carefully in half. What do you find? Largely air. When you pour on the milk, it seeps into the Krispie, displaces the air, and creates a sound. Whether you get a snap, a crackle or a pop depends on how quickly the air is escaping or, conversely, how fast the milk is getting in. The fresher the Krispie, the greater the explosion is my experience. Of course, once all the air has been displaced by the milk, the show's over and the rest of your breakfast will be conducted in total silence.

Why doesn't the chocolate in chocolate chip cookies melt when the biscuits are being cooked?

It's not the same chocolate as you find in a chocolate bar. It's been fiddled with, or tempered as they call it. They take the chocolate and repeatedly heat and cool it till it eventually takes on a crystalline structure which gives it more stability. It also gives it a glossy shine and a crispy surface and it doesn't melt as readily as ordinary chocolate. Also remember that the cookie dough holds the chocolate bits in place, so even if it did melt the chocolate couldn't actually run anywhere.

I've been chopping **onions**.
Why does that always reduce me to tears?

It's enzymes again! When you cut open an onion you release enzymes called allinases. Now, the smell of onions is caused by the presence of sulfoxides, and on contact with the newly released enzymes these are converted to sulfenic acids, which are unstable and quickly convert into syn-propanethial-S-oxide which is – I'm sure you know this – a volatile gas. This unstable gas meets the water on the surface of your eye and undergoes another change, this time into a mild sulphuric acid.

The nerve endings in the cornea of the eye don't like this at all, understandably, and the eye's protective mechanism kicks in causing the lachrymal glands to release tears. This is not necessarily a good thing because it provides even more moisture to meet yet more volatile gas to produce yet more sulphuric acid when you least need it.

It is said that to avoid the tears, cut onions under running water. Another tip, according to some, is that if you hold a slice of bread between your teeth while chopping onions, you won't cry.

What holds **sugar cubes** together? Do they use some kind of **glue?**

No need, when water can do the job perfectly well. When sugar cubes are being formed in the factory, the tiny individual sugar crystals are pressed together under a controlled level of humidity. This controlled water content means that the surfaces of the crystals dissolve slightly in a small amount of water, forming a syrupy solution. Obviously, if too much water is present, the crystals would completely dissolve. But if they get it right, when the crystals are pressed this syrupy solution can flow between the crystals sticking them all together. It's a bit like a brick wall with the 'bricks' of sugar crystals being held together by the syrup acting as mortar. Even when the syrup dries out, the resultant mass of sugar keeps the crystals bonded together.

I've **noticed** that
when you
sprinkle sugar on a **bowl**
of **strawberries,**
after a while you end up with
strawberry juice in the **bottom**
of the **bowl.**
Where's that **come** from?

You have to understand about osmosis, which is defined as the movement of solvent along a concentration gradient across a semi-permeable membrane from the dilute to the more concentrated solution. In your bowl of sugared strawberries, this is precisely the set-up you have created. The semi-permeable membrane is the cell membranes of the strawberry. The weaker of the two sugar solutions is the one inside the strawberry, and the stronger is the one outside which contains our added sugar. The water in the strawberry moves, by osmosis, into the stronger solution in the bowl until the concentrations of the two solutions are in balance.

I've read that **oily-fish**
eaters **have** greater
brains. True?

Possibly. The brain is very rich in decosahexaenoic acid (DHA), a fatty acid which the body can produce, but not very

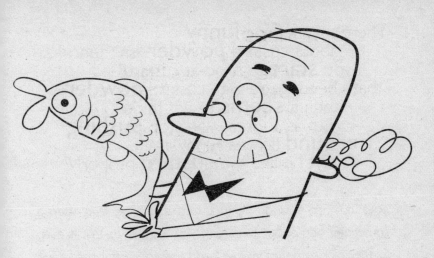

efficiently. The best source for DHA is diet. It is found in meat and eggs and in particularly high levels in fish. Oily fish (like mackerel, sardines, herring and tuna) are very high in DHA, whereas white fish (like cod, plaice and monkfish) have high levels of DHA only in their livers.

It has been reported that DHA improves eyesight, blood circulation and skin, and alleviates rheumatoid arthritis. There is also evidence to suggest that it increases learning ability and visual awareness. Rats fed on DHA-rich diets learn to escape from mazes faster than those deprived of it; experiments on primates revealed similar results. To sum up – eat oily fish!

There's something **funny**
about custard **powder**, isn't there?
If you **start** to prod at custard
that's been made with custard **powder**,
then it **seems** to get thicker.
But as soon as you stop **pushing**
it **around** it goes runny again.
Didn't I make the **custard** properly?

Your custard sounds perfect, and it is true that there's something odd about custard. If you wanted to test it even further, pick some up in your hand and squeeze it hard. You'll find you can make it hard enough to form a ball. But don't try and throw it because as soon as the pressure is released, the custard goes runny again. You have produced what is called a dilatant liquid in which solid particles are suspended in water, which fills the gaps between them. If you stir gently, the custard flows because the suspended particles move around easily. But once you start to apply pressure, the water is forced out of the space between the suspended particles, they start to rub up against each other, and the resulting friction gives the feeling of a thicker liquid – but only for as long as you apply the pressure.

Quicksands are also dilatant liquids, which is why the harder you struggle to get out, the tougher it becomes. Experts say that if you keep still you will float to the top – perhaps it's best to experiment with custard powder rather than quicksand if dilatant liquids interest you.

What about Silly Putty?
It's supposed to be **soft,**
and if you **squeeze** it in your hand
it feels soft. So if it's as soft as it **looks,**
why doesn't it **dent** when you
hit it with a **hammer?**

We're talking custard again, or at least we're still in the area of dilatants. Silly Putty has an interesting history. Its proper name is 'Dow Corning 3179 Dilatant Compound' and it was first discovered in 1943 by the scientist James Wright, who was trying to discover a way of making artificial rubber. He was experimenting with mixing boracic acid and silicone oil and he came up with something which would stretch and bounce further than natural rubber. The trouble was, no one could think of a use for it. But when an advertising executive heard about it some years later, he saw its potential as a toy and soon after that Silly Putty became a sensation. It was even used by the crew of space mission *Apollo 8* to prevent tools from flying around their spacecraft.

Now, to answer your question: Silly Putty starts with a polymer called polydimethylsiloxane, and polymers consist of long chains of molecules – think spaghetti. Adding boracic acid causes the molecules to attach to one another at various points, turning the liquid polymer into a solid compound. But in Silly Putty the links are not all that tight so the Putty is not brittle and can't be broken with a hammer, but gentle pressure will cause it to yield.

How does **superglue** work, and **why** doesn't it **stick** to the inside of its own **tube?**

Most glues work by evaporation of the solvent in which the 'sticky' bit is contained. Superglue, however, is a cyanoacrylate resin which needs hydroxyl ions to activate it – the usual source is water. Because there's no water in a tube of superglue, it will never set. That's why those tubes are so tightly sealed – it's to keep moisture out.

Those tubes are
never quite full,
are they?
Are they selling me **short?**

No, they're doing you a favour. While water activates a super-glue resin, oxygen inhibits it, and leaving a little air in the tube does that job quite nicely. Remember, superglues are looking for warmth and moisture in order to carry out their sticky work. That is why they love your skin so much. Take care.

What makes **cling film
stick** to itself?
What **kind** of glue
makes that work?

No kind of glue at all. Cling film works because of a static electricity charge which is given to the cling film as you pull it off the roll. This highly charged piece of plastic will be attracted to anything which is both uncharged and an insulator, which is why it seems to work best when covering a plastic box. If you try it on a metal saucepan, it will work less well because the static charge is dissipated. You may notice that if you remove a length of cling film from the roll, leave it lying around for a while, and then try and use it, nothing happens. It's simply lost its electrical charge.

I **suppose** both pen and printer ink must **contain** glue, because **how else** would the ink stick to the **paper?**

To some extent, but inks and paints contain pigments which are chemical compounds that do not dissolve in water or oily liquids. One of the most common natural pigments is titanium dioxide – a white substance which is used in everything from emulsion paint to confectionery.

Writing ink is a simple mixture of a very finely ground pigment, a suspending agent and some sort of gum or adhesive that fixes the pigment to the paper. As you write on the paper, the pigment and suspending agent are drawn into the fibres and as long as the paper is not too absorbent, like blotting paper, the ink will stay pretty much where you put it. Then the suspending agent evaporates leaving the pigment trapped in and on the paper.

I've **noticed** that if you **press** Silly Putty on to some newsprint, it will lift the **print** off the **page,** like a rubber eraser when you're **trying** to rub out pencil **writing.** **How does** this work?

Paper, which is made of fibres, consists of bumps and ridges. When you write on the paper, some graphite from the pencil tip is worn off as the graphite molecules from the pencil interlock with the bumps and ridges of the paper molecules. As the rubber is dragged over the page, every graphite molecule that it comes into contact with makes a better bond with the rubber than with the paper, and so the graphite is lifted off the paper. The eraser 'dust' that is left over is simply worn off eraser with graphite stuck to it. A wet rubber won't work because the water molecules lock into the eraser molecules and prevent the graphite from bonding. All you'll get is smears.

If water is made up of **hydrogen** and **oxygen**, why does it put **fires** out? Shouldn't it **burn** instead?

The quick answer is that it's 'burnt' already. The process of burning is what occurs when something combines with oxygen. In the case of water, the hydrogen has already combined with the oxygen and so the 'burning' has already taken place. You could say that water was the 'ash' left over after the fire and once burnt it won't burn again.

It's true that hydrogen is flammable, but oxygen is *not*. If you were to apply a match to a stream of oxygen, the match would burn faster but the oxygen itself wouldn't.

I know you should **NEVER** do this,
but if you've got a chip pan on **fire**
and **you try** to put the fire out
with water, there **seems** to be
some **kind** of explosion.
What's **going on** there?

First, remember that oil floats on water, and that the temperature of the fat will be far above that of the water, and much
higher than 100°C which is the temperature at which water
boils. So, the fat catches fire and you throw on the water. The
water then tries to sink and as it meets the hot fat it starts to
boil. The bubbles of steam quickly rise through the hot fat and
burst violently into the open air carrying with them oily
droplets. If these should find the flame, then they will ignite too.

Trying to extinguish a chip pan fire with water is one of the
most dangerous things you can do in a kitchen. Instead, place a
damp cloth over the pan to deprive the flames of oxygen, and
they quickly go out.

We've got lots of **clothes drying**
in our kitchen. I was wondering
how that works **because** the kitchen's
never hot enough to **boil** the water
and **make it** evaporate.

Although the temperature of the clothes is not raised to
boiling point, water does still escape from the clothes into the

air. Any single water molecule on the clothes is attracted to the other water molecules, and also to the molecules that make up the clothing. The water molecule finds itself in a 'sticky' environment which means it is hard for it to escape into the air, but it does have enough energy to move around, swapping positions with other water molecules.

Heat can be thought of as energy that molecules possess, and the more heat, the greater the amount of energy they have, and the easier they find it to overcome the stickiness of their surroundings. At a room temperature of 20°C, for example, some water molecules will have enough energy to overcome the forces of attraction from the other molecules, and will be able to escape completely from the clothes into the air. As this process continues there will be less and less water remaining on the clothes until eventually they are dry.

If the temperature is higher than 20°C, then more molecules will have enough energy to evaporate, and therefore the clothes will dry even quicker. Some of the water molecules that have escaped from the surface of the clothes into the air may fall back down on to the clothes and get stuck again, which is why putting clothes out on a windy day helps to speed up the drying process because the wind blows away the water molecules that have escaped from the surface of the clothes, so they are less likely to get stuck again.

I had a **disaster**
with a wool sweater
that I washed in **water**
that was **too hot.**
It shrank!
Why did it do that?

Because wool is made of scaly fibres. If you were to look at a strand of wool under a powerful microscope, you'd see that its surface looks like the surface of a stack of polystyrene cups. Under normal conditions these rough surfaces can't move much against one another, but if immersed in warm water then the scales on the fibres can ride over one another. But when the wool dries they can't get back again – your sweater has then shrunk and the wool has become matted because the surfaces have locked together.

Wool is funny stuff. The outside of wool hates water and repels it (hydrophobic), but the inside of a wool fibre is hollow and absorbs water (hydrophilic). If you wet a wool sweater, it will repel water until it reaches a point where it gives up and instead starts to soak it up. If this absorbed water is then heated, the fibres become slippery and that's when the shrinkage starts and it can't be reversed. Once a wool sweater has shrunk, that's it.

How many **sheep** does it take to **produce** enough **wool** for a sweater?

If you reckon an average sweater weighs 250g, and you get 5kg of wool from an average sheep (only 65 per cent of which is usable because the rest will be too filthy), then you could expect to get fourteen sweaters from one sheep's fleece.

Why does an iron work better when **it's hot**, and even **better** if it's a steam iron?

When you iron a shirt, you are trying to flatten all the fibres and encourage them to stay that way, and heat makes them that bit easier to flatten (although too much heat, of course, will burn them). Moisture helps to soften up the fibres too, which is why steam helps, although for the best results a shirt fresh from the washing machine and still slightly damp will give the best results.

The best way to iron a shirt (if it's not still damp) is to spray it lightly all over with water and roll it up into a ball for a few minutes. This softens the fibres making the iron move across the fabric more easily. Also, turn the shirt inside-out so that you are ironing on the 'wrong' side of the fabric. There are two reasons for this – (1) any dirt on the iron that might be transferred will be less noticeable if it's on the inside of the shirt, and (2) by ironing the inside you are 'pushing out' the fabric 'away from the body' so it hangs better.

If I've **washed** my hands in **soapy** water and then dry them on a **towel**, is it the **water** or the soap that's making the towel **wet?**

Bubbles are a mixture of water and soap molecules. Water molecules have two atoms of hydrogen and one of oxygen. In liquid water there are certain forces between these molecules called hydrogen bonds. They are quite strong forces and they allow water molecules to hold together and remain as a liquid, as long as the temperature isn't raised above 100°C – boiling point.

Soap molecules are rather different. They are essentially long molecules with two ends. One end is called 'polar', which means that it's charged at that end. Charged molecules interact with water molecules (and therefore dissolve) very easily. The rest of the molecule is a long 'tail' which is not polar. Non-polar molecules don't dissolve in water so well – a good example is oil.

So with a soap molecule, one end 'likes' water and the other doesn't. This means that if you put soap in water, the soap molecules' preferred place is on the surface, with their non-polar ends sticking out into the air (so they don't have to interact with the water).

Now we get to the bubbles. The non-polar ends of the soap molecules 'like' air, so they are sticking out on the outer surface of the bubble. Their polar ends are in contact with water molecules, which they love. And on the inside of the bubble, once again we have these non-polar ends, the water-haters, sticking in towards the centre of the bubble (where there is also air). This means that the non-polar ends form an inner and outer 'shell' to the bubble, and between them the polar ends stick to a single layer of water molecules.

So, bubbles are made of both water and soap molecules and if you rub them on to a towel, you're going to get a bit of both.

Will **hands** washed in only water dry **faster** than hands **that** are soapy?

If you don't rinse your hands before putting them under a blower, then they will probably take longer to dry. This is because the detergent will find its way to the surface of the film of water on your skin and form a protective layer. This prevents the water evaporating quickly. This is one of the reasons why bubbles made only of water and no soap don't last very long; it's simply because the water evaporates too quickly. Also, the surface tension of water on its own is higher than that of soapy water, and this tries to pull the water down into a droplet.

We've got **soaps** of all colours
in our house,
but I notice the **bubbles**
are always **white.**
Why **don't** the bubbles
pick up the **colour** of the soap?

What you're seeing hasn't got much to do with the colour of the soap. The effect is created by reflection and diffraction of light by the thin film of water that is sandwiched between two layers of soap. So, you're getting all the colours of the rainbow, which you can often see swirling on the surface, and some reflected white light as well.

How does hair-gel work?
You squeeze it out of the **tube**
and it's a liquid,
but when you **rub it** on your hair
it turns **solid.**
What kind of **physical** change is that?

Not as clever a change as you might think! If you take the time to read the list of ingredients on the tube you'll see that by far the biggest constituent is water. When this evaporates it leaves behind the other ingredients, which I'll list because I suspect you'd never guess exactly what you plaster on your hair once a day. There's polyvinylpyrrolidone, which is a resin and is the bit that fixes the hair in place; there's the gel agent which thickens the water and makes it into gel; also a fragrance solubilizer (PEG-40 hydrogenated castor oil) which enables the oil-soluble fragrance to dissolve in the water. Then you've got preservatives, and neutralizers to maintain the clarity of the gel. It's the resin that holds the hair in place.

It would be **great** if you **could** have a hair **gel** and shampoo all in the **same** bottle. **Given** the way they both work, is that ever going to be **possible?**

Yes, although the makers would probably rather you bought two bottles instead of just one, which might explain why you don't find it in the shops. Of course, you can get conditioner and shampoo in the same bottle, and a shampoo/gel mixture might be able to work in the same way. What happens is that the shampoo contains surfactants which considerably lower the surface tension of liquids – water, in this case – which makes for easier bubble formation. This is what happens when we set to and work up a lather. While this is happening, the conditioning agents are being held in suspension in the lather, and are left behind when you rinse the shampoo surfactant out of your hair. It's possible that gel could be left behind rather than conditioner.

I **notice** that if I pour Dettol (a disinfectant) into **water,** it goes cloudy. And then I **remembered** that there are a **couple** of alcoholic drinks, Ouzo and Ricard, which do **the same.** I've heard of **colloids.** Is that what's **happening** here?

Colloidal particles are so small and light that they don't settle in water because there is enough energy in the movement of the water to stop them. But that's *not* what's happening in the case of Dettol. In Dettol, the main constituent is pine oil dissolved in alcohol. Now, pine oil is very happy to lurk in alcohol but is not so happy in water, in which it will not dissolve. So, as you add water to the Dettol, the alcohol becomes more dilute and its capacity for holding the pine oil decreases; the oil 'drops out' into the water and appears as a fine suspension, which is what makes the water go cloudy.

The same thing happens with Ouzo and Ricard which rely on compounds called terpenes for their distinctive flavour. Like pine oil, these will happily dissolve in the strong alcohol base of these drinks (usually 40 per cent) but as you add the water they too 'drop' out and turn the drink cloudy.

Some of the **bricks** they used
to build our house had a **hollow** side
and a **flat side.** Why is that?

We're talking frogs, which is what the hollow sides of bricks are properly called. When a wall is built, the bricks should always be laid with the frog upwards. The frog does several things: it reduces the weight of the wall because you will have less brick material, but it also strengthens the wall because the mortar actually goes into the brick rather than just lying on its surface. You don't want to have two frogs facing each other because you would use up lots of mortar and the brick would not be as strong – it would have a weak centre.

The reason bricks are laid frong upwards is to ensure an even load. You have the brick, then a layer of mortar which completely fills the frog, then another brick. If the frog is downwards you cannot guarantee that the frog is completely filled – it may contain an air gap and then the load will not be evenly spread throughout the brick – instead the load will be channelled down each side.

I've noticed that when my **builder** uses
a **chisel** and hammer, **sparks** usually fly.
Is this **static** electricity, or **what?**

Sadly, it's bits of his chisel he's losing and what you're seeing is a fragment of molten metal flying through the air. The friction caused by the strike makes the metal heat up so much that it becomes molten – this is the spark that you see – and it glows because it is so hot. In fact, it is burning because it's combining with the oxygen in the air to produce iron oxide.

How do **sparklers** work, then?

Hot oxide particles, again. A sparkler is covered in a silicon and oxidant paste impregnated with iron filings. When you light a sparkler, high temperatures are produced which burn the iron filings (producing iron oxide) which fly off in all directions. Although they're very hot, they're also very small, which is why you only feel a tingle and not a burn.

Why is it that **sparklers** are quite **harmless** fireworks, but rockets **blast** high into the **air** and **Catherine** wheels spin violently?

Fireworks are like small space rockets and work in the same way: they burn lots of fuel, usually gunpowder, in a confined space, which blasts out of the bottom in what is effectively a controlled explosion. Gunpowder is a mixture of potassium nitrate, charcoal and sulphur.

Some fireworks are fired, like bullets from a gun, from a tube called a mortar, and when they are very high they explode into lots of colours. The bit of the firework that gives the explosion of colour is called the shell, which contains 'stars' or little balls of chemicals. A bright white firework might contain some magnesium stars, while a red one might contain strontium. The shell also contains a 'charge'. This is fixed in the middle of all the stars and when the charge explodes it throw these stars all over the sky to burn in bright colours.

A Catherine wheel is just a rocket firework that can't get away from the ground because it's usually pinned through the middle. Think of a jet aircraft fixed to the earth by a big, strong rope tied to the top of the Eiffel Tower. If it tried to fly in one direction, it wouldn't get very far because the rope (unbreakable, of course) stops it. The only way to dissipate the energy produced by its engine would be to go round and round in circles on the end of the rope. It's the same with Catherine wheels.

Firework chemistry is more complex than it might appear. The colour the firework produces can be a result of either incandescence (light produced by heat) which varies with temperature – red heat is cooler than white heat – or luminescence, which results when energy is absorbed by an atom's electron making it excited and unstable. When the electron shakes off this surplus energy, it emits a photon of light and the energy of the photon determines the colour of the light you see:

Red	strontium salts, lithium carbonate
Orange	calcium salts
Gold	hot iron or charcoal
Yellow	sodium
Bright white	hot magnesium or aluminium
Green	barium
Blue	copper
Purple	mixtures of strontium and copper
Silver	burning aluminium, titanium or magnesium powder

Naturally occurring **colours** are interesting. Why do we get **different** coloured diamonds? I thought **diamonds** were carbon **crystal** structures **and** nothing else. But some have **a tint** to them. **Where** does that **come** from?

Diamonds are carbon crystal structures, and can carry a slight tint for various reasons. First, there can be traces of substances within them; for example, nitrogen gives them a yellow colour, boron blue. But if the lattice structure of the crystal is deformed in any way, that can lead to a colouring without there being any 'impurity' present. This kind of deformation leads to brown, pink or red diamonds, all of which are rare.

If diamonds are just **crystals** and so is **salt** (sodium chloride), why are **diamonds** hard while **salt** is soft?

It's all down to the bonds between atoms and molecules, of which there are two types: covalent bonds and ionic bonds. In diamonds, any single carbon atom is bonded to four others by covalent bonds in which there are shared pairs of electrons. This is a strong kind of bond, as beefy as they come.

In salt, consisting of ions of sodium and chlorine which are positively charged (sodium) and negatively charged (chlorine – called a chloride ion), the ions are attracted to one another and hold together by attractive electrostatic forces called ionic bonds. Ionic bonds involve one ion giving an electron to another and these bonds aren't as strong as covalent bonds. This makes the diamond the winner in the strength stakes.

Is there **anything** harder than diamond?

Diamond is still the hardest substance known to science, but a group of American researchers say that they have produced a composite material which contains crystals of carbon nitride, a substance which scientists believe stands a good chance of being even harder.

The search for super-hard materials started in earnest in the late 1980s when an American scientist came up with an equation for calculating a substance's hardness. This showed that beta-carbon nitride (beta-C^3N^4) should be a cut above the rest.

Non-crystalline carbon nitride, which is a blue-grey substance, is easy to make in the laboratory, but the super-hard crystals have proved extremely difficult to obtain. The researchers alternated thin layers of carbon nitride and titanium nitride at room temperature using a process called 'magnetron sputtering' in which molecules of gas are fired at a solid target. The gas molecules knock atoms off the surface of the target, combine with them chemically, bounce off and are deposited on a nearby surface. The team decided to fire nitrogen molecules at a target that was coated half with carbon and half with titanium so that as the target rotated the nitrogen molecules hit the two materials alternately, giving successive layers of titanium nitride and carbon nitride on a surface held next to the target. Titanium nitride and non-crystalline carbon nitride are both hard substances, but the composite material, which had a faint pinkish hue, turned out to be twice as hard as either. But, it still wasn't as hard as diamond.

However, the search is still on because something harder (and cheaper) than diamond would have a wide range of uses. Super-hard materials could be used to cut steel, which diamond cannot do because it burns when it gets hot. It is also impossible to coat metals with a thin layer of diamond. Mechanical components such as gears and bearings coated with beta-carbon nitride would last much longer than normal parts, and could be used in devices where liquid lubricants are unsuitable. A thin layer of the beta-carbon nitride could also be used to protect the surface of computer discs.

So, if **diamond** is still the hardest substance **known** to man, how **come** we can cut it?

Because, believe it or not, diamonds have a grain, like wood, and if attacked in the line of the grain will split cleanly. If you want to cut a diamond with the grain you can use an 'axe', a sharp metal blade which is hit (carefully) with a hammer. If you want to cut across the grain you use a saw which is made from a paper-thin disc of phosphor bronze impregnated with diamond dust, revolving at up to 10,000rpm. As the saw cuts it impregnates itself with dust from the diamond it is cutting, so maintaining its own sharpness. Even so, a sizeable diamond can take up to two weeks to cut right through.

You **could cut** a lump of glass in the same way, **but why** wouldn't you get the same **sparkle** that **you get** from a diamond?

Diamonds have a much higher refractive index than glass, so even if they are cut in exactly the same way the diamond will sparkle more as it splits up light into its colours much better than glass.

The secret of a diamond's beauty is the way it reflects light, and the cutter must shape the stone in such a way as to admit light through the top of the stone, so it will bounce around inside and come back out at the top. In this way the maximum amount of light is reflected and the diamond sparkles.

By the beginning of the twentieth century the art of cutting diamonds had been so refined that a precise mathematical formula was developed. It called for most diamonds to be cut with fifty-eight facets, each placed at a precise angle to the next.

By the way, the 'cut' of a diamond is not the same as its shape. The shape is a matter of personal preference and doesn't affect the value, but the 'cut' does. A good cut gives you maximum scintillation and dispersion of light through the diamond. This only happens when a diamond is cut to good proportions, and light is reflected from one facet to another and then dispersed through the stone. If light escapes as it travels through the diamond, the sparkle is largely lost.

Coal is made of **carbon** and so are diamonds. Are **diamonds** and **coal** the **same thing?**

It is a great myth that diamonds and coal are related in any other way than coal is a mineral rich in carbon, and diamonds are made of carbon too. That's as far as it goes.

Diamonds are formed under very high pressure in the Earth's magma, well below the crust, and are often associated with volcanic pipes in South Africa where magma from the interior of the Earth 'escaped' through pipes leading off from the centre core of a volcano a very long time ago. Over centuries and centuries this magma was subjected to incredible pressure as more layers were deposited on top of it. Eventually this pressure compressed the magma into very durable and highly valuable diamond.

Coal, on the other hand, is a general name for firm brittle carbonaceous rocks which are derived from debris such as decaying trees, leaves and vegetation. Coal was initially deposited a long time ago as peat, but burial and increase in temperatures at depth resulted in physical and chemical changes producing coal.

So, it's true that coal and diamond both contain carbon and were formed under very high pressure but they are completely different substances. It is also true that coal is too impure a form of carbon to result in a perfect diamond, even if incredible pressure was applied to it.

Since **coal** burns, **does that** mean diamonds will burn **too?**

You can burn a diamond if you get it hot enough. Coal starts to burn when it gets to 400°C, but diamond won't ignite until it reaches about 800°C.

The reason for this difference is the way that the atoms of carbon are put together in the two materials. Coal is made from the remains of very old plants, and the carbon atoms are arranged in a rather untidy way, without any regular pattern. Think of the atoms as a pile of Lego bricks which you have just tipped out of the box: you can move them apart very easily, and join them to other bricks without much difficulty.

Now put the bricks together so that each one is joined to four other bricks around it. As you add more bricks, you get a very strong solid shape which is rigid and needs a lot of effort by you to pull it apart again. This is a 'Lego diamond'. In a real diamond the bricks are carbon atoms, but the pattern is the same. Because of the regular way that the carbon atoms join together in a diamond, it is very hard – you can't push the atoms around easily.

When something burns, the atoms have to be separated from each other, and this takes a lot more energy for diamond than it does for the untidy heap of atoms in coal, which is why you have to get it much hotter to set fire to it.

I **always** think my voice sounds deeper and **richer,** and my **singing** is certainly better, in the bathroom. **Why** is this?

People listening on the outside may not agree! What's happening is the bathroom is being very kind to you. Unlike other rooms, bathrooms are full of hard, reflecting surfaces. Think of those shiny walls, hard basins and baths, and floors

without carpets – perfect surfaces for reflecting high frequencies. In an ordinary room, with soft furnishings, the high frequency sounds – which is what most singing consists of – are absorbed but the lower frequencies survive. To prove the point, the bass guitar playing on next door's stereo is heard more easily through the wall than the cymbal of the accompanying drummer. So, if you're singing in a bathroom you've got all those high frequencies coming back at you which you wouldn't get if you sang in a sitting room.

You've also got to consider resonance. Objects have a certain frequency that they 'prefer' to vibrate at. In a bathroom it may mean that certain frequencies appear louder than others because they are at the same resonant frequency as the walls and other surfaces. If the resonant frequency is pleasing musically, then you will obviously like the sound you hear and you'll start to think your singing is really good.

8

Got
that
Feeling?

Curries, Sherbet and
Falling in Love

Curry can sometimes be a pain. What is the **chemical** that makes the **burning sensation?** Since it feels like fire, is water the best **thing** to put it **out?**

The chemical component of peppers which makes them hot is the alkaloid capsaicin, and it's one of five components that have different effects on the mouth. Three give 'rapid bite sensations' in the back of the palate and throat, and the other two give 'long, low-intensity bite' on the tongue and mid-

palate. How much burn you get from your peppers is due to differences in the proportions of these five. Capsaicin is a general irritant and so can 'burn' skin that is already damaged by cuts or abrasions. The body's response to capsaicin is a defensive one – pain, watery eyes, runny nose.

Should you drink water to try and cool your mouth if you're hit by a really hot curry? No, because capsaicin is oil-soluble, so drinking water will not do the trick of getting rid of the burning sensation. Try drinking milk or yogurt, both of which contain oil in which the capsaicin will dissolve and be digested. Why do you think those kind people have put raita on the menu?

How does **sherbet** work? I **love** the fizz.

Would you love it as much if I told you it was a form of pain, which is all down to enzyme action? When sherbet dissolves in the mouth and erupts into a tingling, foaming sensation, it's the same process as when a carbonated fizzy drink hits your tongue. First, a mild acid is formed in the mouth, or on the coating of the tongue, by the action of an enzyme in the saliva which creates a weak carbonic acid. The acids in sherbet, which are citric and tartaric acid, work in the same way. Sherbet lovers could be classified as serious masochists. Acid-suckers, we now call them.

Sometimes, when I bite hard on **a Polo** mint, I get a spark. **What** is that?

Is your name Jaws, by any chance? There are actually two sources of light in polo mints. The first comes from the sugar molecules, and the second is the minty wintergreen flavouring, a molecule called methyl salicylate.

When you bite into your mint, as your teeth crack its sugar molecules, positive and negative charges are separated. When the charge difference is large enough, negatively charged electrons jump across the crack, bumping into nitrogen atoms on the way. The nitrogen is there because it is one of the components of air. The collision between the electrons and the nitrogen causes the nitrogen to emit a very faint blueish light. If you want to see this effect without risking your teeth, you need to wait in a darkened room for a good fifteen minutes to allow your eyes to come to maximum sensitivity. Then simply strike two sugar cubes against each other and you'll see the same blueish effect.

Why do **mints** make your **mouth** feel cool?

The taste of a mint is picked up by the four types of taste buds on the tongue – sweet, bitter, salty and sour – and by smell sensors in the nose. The peppermint flavouring that gives mint its taste is like an activator that sends the 'mmmm … minty' message to the brain.

But the cooling effect of a mint isn't anything to do with the peppermint flavouring, or the senses of taste and smell. Cool sensations are sent to the brain along different nerves, which are usually activated by a drop in temperature. One chemical that can trigger these 'cool' nerves into life is called menthol, which is an important ingredient in mints. This is what makes your mouth feel cool, but there's no real drop in temperature. It's an illusion.

Why does rubbing an injury make it hurt less?

Signals travel to the brain via nerve cells or neurons, but some neurons transmit signals faster than others. Pain signals travel more slowly than some others, and when you rub your injury – which is just a way of causing it to heat up – that signal goes along a different set of sensory neurons from the pain signal and has a faster transmission rate. So the message from the rubbing (or the heat) gets to the brain before the pain message does, and it appears that the pain is lessened.

Why
do we itch?

Itching is an early warning system to alert you to the fact that your body has come into contact with a noxious substance. At the ends of our nerve fibres are tiny organs which are able to receive messages and send them on to the brain. Some of these organs might be sensitive to heat, light, pressure or pain. It is when the 'pain organs' are stimulated that we feel the itching sensation.

Chemically speaking, when we rub an itch we release from our 'mast cells' (which are a type of white cell involved in allergic reactions) a substance called histamine. Histamine binds to the receptors on the local nerve endings, causing the sensation of itching.

Itching can be stimulated by various irritants. People with an allergy over-produce histamine in response to contact with a substance that may not affect another person at all.

What an 'itch' really is, in a physiological sense, is not well understood, but it may be a special kind of pain sensation only felt under the influence of precise stimuli. Some say it is nothing to do with pain at all, and is a sensation in its own right with its own mechanisms.

Why does a **wound**
itch when it's healing?

When cells are damaged by cuts, chemical agents or bacteria, then this injury causes an inflammation response which is part of the body's defence mechanism. It usually has any of four symptoms: redness, pain, heat and swelling. Inflammation is the

ITCH

ITCH

ITCH

body's attempt to destroy microbes, toxins or foreign material at the site of the injury so that it doesn't spread to neighbouring tissues, and also to prepare the site for tissue repair.

The actual itch that we feel as the wound heals is caused by the growth of new cells underneath the old scab. As new skin cells form a new layer of skin, then the scab becomes more tightly stretched. This can make it feel itchy. Nerve cells will also be developing under the scab, and as these start to receive and send messages, they will create that itchy feeling.

Why do we **scratch**,
and what
gives us **sudden** itches?

First, we have to understand the 'itch and tickle sensation' which is caused by the mechano-receptive nerve endings found in the upper layers of the skin.

A mechano-receptor is a cell, or part of a cell, that contains structures sensitive to stimulation by twisting or bending. They are very similar to the cells responsible for that familiar slow type of ache, and so it has been assumed that itching is just another type of pain.

What happens when these nerve endings are stimulated is the release of a neurotransmitter (called substance P) which causes blood vessels to dilate allowing more blood to the area of irritation – this causes your skin to go pink. It also

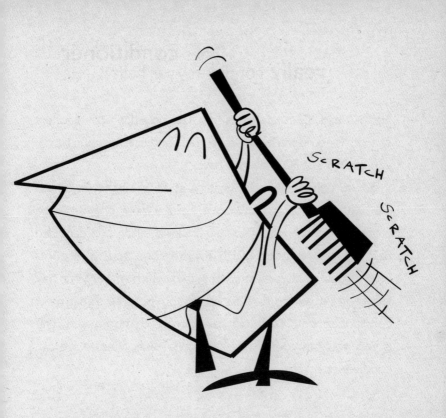

SCRATCH

SCRATCH

activates the mast cells, which are part of the body's allergic response system, and these cells release histamine which causes the blood vessels to dilate even more and makes the irritated area swell. This is what produces the itch and tickle sensation.

The scratch reflex is a powerful one and is caused by a spinal cord reflex that locates the area of the itch and directs your hand to it. The scratching relieves the itch either by removing the irritant or by suppressing the itch signals in the spinal cord if the scratching is hard enough to cause pain.

Does **conditioner** **really** condition your hair?

Hair is dead. It's as simple as that. The only thing conditioners can do is temporarily improve the condition of your hair – you can't give it life it doesn't have.

So do conditioners do anything at all? Well, the more you play with your hair – blow-dry it, perm it, even brush it – the more damaged it gets. If you can imagine taking one damaged hair and looking at it through a microscope, you will see lots of protruding scales all stuck together, which makes your hair feel rough; it will be hard to brush, and look dull. What most conditioners do is coat your hair to make it smooth again. This is why you can comb your hair more easily after you've put conditioner on it.

Why do we **yawn**, and **why** is it catching?

There are three theories and they all conflict.

First, the physiological theory. We yawn to get more oxygen into our systems, or to rid ourselves of excess carbon dioxide. Yawning is therefore contagious because everyone in any given room is likely to be short of fresh air at the same time.

Secondly, the boredom theory. If everyone finds something boring, then they will yawn. However, this doesn't explain why

we yawn when bored unless it's just a social signal to a third party that they're dull.

Finally, the evolutionary theory. This says we yawn to display our teeth and to show that we can get nasty if we have to. Yawning once acted as a warning to others, but has lost its aggressive meaning as we have become more civilized.

Convinced by any of those? A scientist called Dr Provine has done several experiments on yawning and concludes that the first theory is incorrect. He taped closed the mouths of his subjects so that they couldn't yawn by extending their mouths, but could only take air in through their noses. His subjects said that they did not feel 'satisfied' with this method of yawning, even though they could still take oxygen in. He also pumped pure oxygen into his subjects, only to find their frequency of yawning did not change. This would suggest that it is not the urge for more oxygen that caused the yawning.

As for the boring hypothesis, he found that considerably more subjects yawned while watching a thirty-minute test pattern than while watching thirty minutes of rock videos. But did they yawn for psychological reasons (they were bored) or for physiological reasons (boredom made them sleepy)?

Provine did find, however, that most yawns occur about an hour before sleep and an hour after waking. There is also an unmistakable link between stretching and yawning.

But now there's a new theory which says that yawning is a way of stimulating the flow of lymph through our facial muscles. Lymph is a fluid which flows through the body's lymphatic system helping to fight infection and disease. In order to move around the body, lymph needs the skeleton to be in motion because it doesn't circulate in the same way as blood. It's been suggested that this is why we enjoy stretching first thing in a morning, after sleep. Yawning could be a way of getting the lymph moving in our faces and necks.

Why is it contagious? There's no real answer to that, but yawning definitely is. Simply writing this made me yawn, and I bet you felt like yawning too – thought not, I hope, through boredom.

Why do we use our **elbows** to test the **temperature** of babies' bath **water?**

It would seem to make more sense to use your hands, but although there are more nerve endings in the hands than in the elbow, the skin in these regions is usually quite thick and shields our skin temperature receptors from the heat we are trying to test. Also, our hands may have become conditioned to touching hot things and so will not be a good indicator. So using a region of the body which has thinner skin is more effective. The elbow is probably used because it's handy.

Why do we crave **chocolate?**

Chocolate has chemical properties that cause pleasure. That's it! There are large amounts of phenylethylamine, which is also present in our bodies and released during sexual arousal, heightening sensation and raising the heartrate. Chocolate also contains methylxantine and theobromine, which have a similar effect to caffeine. And if that isn't enough, it is solid at room temperature but melts at just below body temperature which makes it near perfect.

Why
do we **laugh?**

Laughter is odd because it can show either happiness, nervousness, embarrassment or disappointment. It can be used to relax people (muscles relax throughout the entire body when we laugh), and it can also be used to exclude people by laughing at them. It may also be a way of demonstrating control – when the boss makes a joke everyone laughs.

But exactly 'why' we do it still baffles scientists. Behavioural psychologists will argue that laughter is not quite the sophisticated conscious process that you might expect, but a primitive response to our environment. Laughter may strengthen social bonds since it is an outward sign that we are feeling comfortable in present company – sharing a joke is essentially a form of social bonding. Laughter is also a natural form of relaxation and we have all felt 'weak' from laughing too hard. Laughing has the opposite effect on our body to the classic 'fight-or-flight' response.

In a tense situation laughter may be a nervous reaction, a way to diffuse the potential threat of the confrontation, as in the nervous giggle. Laughter may also be linked with power and aggression.

There may be many reasons for a chuckle, but there's still no final punch line in laughter research.

What makes us sleepy?

Sleep is one of the most common human activities, and yet little is understood about how it works and what triggers it.

The pineal gland located at the base of the brain might be important because this is where a chemical called melatonin is produced. This enters the bloodstream to regulate the cycle of sleeping and waking. Chicks injected with melatonin fall

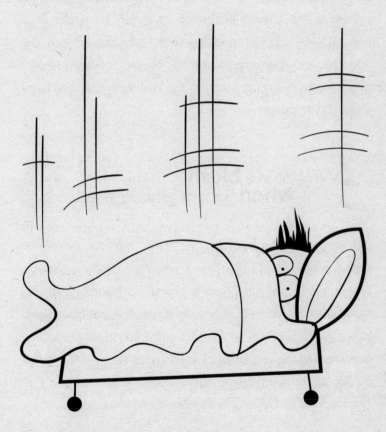

asleep. Until recently no one had ever discovered a *natural* sleep-inducing chemical (although, of course, there are several synthetic drugs that induce sleep) but a group of researchers in California found an increase in the amount of one substance in the cerebrospinal fluid (the fluid that bathes the brain and spinal cord) of sleep-deprived cats. When they injected it into rats, the rats in turn fell fast asleep. This 'sleep' agent is a fatty acid, similar to one of the components in cell membranes, but quite what triggers its release is not known. In the future it may be useful as a natural sleeping pill; pills currently prescribed can be addictive and have unpleasant side-effects not unlike those of a hangover. A drug closer to the real thing may not have these disadvantages.

Why do we **blush when** we are embarrassed?

Blushing is caused by the sympathetic nervous system, a system made up of nerves over which we have no control. No matter how hard you try, you can't stop yourself blushing. In fact, you will only make it worse. Emotions trigger blushing which causes an increase in the blood supply to the face, creating the red-faced look. This returns to normal quite quickly as the sympathetic nervous system relaxes, and the circulation of blood returns to normal.

If the **pupils**
of your eyes dilate
when you are sexually **aroused,**
but contract
in **bright** sunlight,
which takes priority if you
fancy someone
you see sitting on the beach?

There are two different nervous systems at work when it comes to opening and closing the pupils of your eyes: the sympathetic system controls opening or dilation, and the parasympathetic controls constriction or narrowing. The size of the pupil of your eye is always determined by the balance between these two.

So if the beach is very bright, then the parasympathetic system would always attempt to ensure that the eye is closed down to prevent damage to the sensitive retina at the back of the eye. However, in sexual arousal, which is accompanied by increased heartbeat, then the sympathetic system would want those beautiful eyes to be wide open.

In the long run, the sun would win! The parasympathetic nerves which control the constriction of the pupil are more continuously active than the sympathetic nerves that cause dilation of the pupil in states of arousal. So, I would guess that the pupil briefly dilates when you see someone you fancy on the beach, but very rapidly constricts again. Unless, perhaps, the state of arousal was overwhelming.

How **many** nerve endings **does** the **tongue** have?

Are you thinking of taste buds? They're not the same thing. A single taste bud may be connected to more than one nerve ending. Also, some nerve endings detect temperature, others are involved in movement, and some respond to damage by sending pain signals to the brain. There are also cranial nerves

involved, so the tongue is quite a spaghetti junction of nerve endings.

But as a rule of thumb (or tongue) you can reckon that we have a total of 10,000 taste buds not only on our tongue but on our palate and cheeks, and the number of receptors on the end of each bud is between 50 and 150.

Why does **grated** cheese taste **better** than slices of **cheese?**

The taste buds on our tongue and inside our mouth work by a chemical reaction between the food substance and the taste bud. For this chemical reaction to happen, the food substance must be touching the taste bud. With grated cheese, there is much more surface area of cheese compared with the same volume of ungrated cheese. Therefore there is more surface area available to interact with the taste buds, so the taste will be stronger and more varied. This, I might add, is a theory. Test it for yourself.

Sometimes when you eat, your **nose** runs. **Why?**

Your nose might run when you eat for several reasons. If the food you are eating is hot, then the heat inside your mouth will spread up into your nasal cavity. There is always mucus inside your nose, which does a very important job of helping to keep your nose free from germs, and this mucus will become more runny the hotter it is. So eating hot food could make your nose run, because that's where the heat goes. It's not just with high temperature foods that you'd find this effect: spicy foods, like a hot curry, might make your nose run too.

A small amount of nasal mucus is essential for our sense of taste because a lot of what we perceive as our sense of taste is actually smell. People who have lost their sense of smell (like when you have a head cold) complain that food is tasteless and boring. This is because our taste buds are only able to detect four different tastes – sweet, salty, bitter and sour. For the tastes in between we have to rely on the smell receptors in our nose.

Smell becomes stronger when it comes into contact with a moist surface because the odour-producing chemicals can dissolve into the fluid making it easier for the smell detection cells to detect them. This is part of the reason why dogs, with their wet noses, have a much more sensitive sense of smell than humans. So the nose will try to produce a little bit of mucus while we are eating in order to help us smell and taste our food.

What
is the biological **explanation**
for **love?**

If you analyse chocolate, you will find that one of its major 'feel good' ingredients is a chemical called phenylethylamine. This same chemical produced in our pituitary gland is present in our bodies during sexual arousal. It plays a part in heightening sensation and raising the heartrate.

Dopamine is also part of the 'in love feeling' which rushes through the brain making us feel good, helped by norepine-phrine which stimulates production of adrenaline which in turn gets your heart pounding. Most of the feeling of bliss comes from the phenylethylamine. Taken together, these chemicals can sometimes override the brain activity that governs logical thinking – hence, 'madly' in love.

Irrational romantic ideas are thought to be caused by oxytocin, a primary sexual arousal hormone that signals orgasm and feelings of emotional attachment. As you become increasingly aroused, more oxytocin is produced.

Why does time
seem to go **faster**
as we get **older?**

We measure how fast time seems to be passing by relating it to previous experience. The longer we live, the more time we have experienced. At the age of five a week seems longer than it does at twenty simply because a five-year-old has not experienced so much 'time' as a twenty-year-old. Compared to all the time experienced beforehand, a week seems relatively shorter to a twenty-year-old than to a five-year-old, for whom a week is still a significant proportion of their lives.

Why does my **flesh** creep when I touch **cotton wool?**

There are many sensations similar to the one you describe, and they come with different names: they might be called creeping flesh, a shiver running down your spine, goose bumps, or that feeling we describe as 'someone has just walked over my grave'.

Goose bumps are the easiest to explain: they are a leftover from the time when our ancestors were hairier than we are now. Cold air causes the muscles at the base of our skin to stiffen, and this makes the small hairs on our body stand up. If we were hairier, this would allow us to trap more air between the hairs and this would insulate our bodies against the cold. Other animals with hair, and birds, still use this mechanism to protect themselves from cold weather. So, goose bumps are a vestigial trait, which means that our body reacts in that way because it was useful in the past.

You might have noticed that the fur of certain animals stands up not only in response to cold temperatures, but also to threatening noises or sights: think of a cat with its fur standing on end, running away from a dog. This reflex allows them look larger and scarier to their adversaries. In humans, it is possible that when we had more hair we reacted in the same way, which would explain why we get goose bumps when we feel threatened or are uncomfortable.

These reactions will be triggered by different things for different people. Some might sense the noise of nails scratching a blackboard as a threat, while others will be fine. In your case it's cotton wool – and I can't stand the stuff either!

Why do we kiss to show affection?

One of the things that makes us different from other animals is the way in which we communicate with each other. We can do this in a variety of ways, some of which are extremely complex.

One obvious means of communication is language through which we can express complex ideas and feelings, enabling us to be understood by others. It is not clear whether language evolved as we started living in larger groups, or whether language enabled this to happen. Nevertheless it is language that has enabled us to develop distinct cultures and to live in very large groups.

However, humans communicate not only through speech but by using facial expressions, posture and physical inter-action. Some animals also do this, and the larger the social group the more complex the interactions. In social primates, for instance, grooming is an important part of their behaviour, and this serves the function of building relationships.

Although we all have the same ability to express feelings through physical expressions and interactions, the particular way in which we do this will be greatly determined by our cultural environment. So, kissing is a form of communication in which we are telling others how we feel about them, but there is no doubt that the way in which we kiss and the situations in which we think it is acceptable to kiss others will depend greatly on our cultural background.

The lips and the hands are involved in these tactile communications, probably because of the density of sensory nerves in the fingers, lips and even the tongue. All these sensors are involved in sensing the flavour and texture of food and are very sensitive to temperature and texture. Because of their sensitivity, it follows that affectionate bonding has developed using these areas of the body.

What's that **feeling** you get in your stomach when you're on a **rollercoaster?**

First of all, think of the forces acting on your poor insides. On a rollercoaster, as in every other situation, gravity works to try to pull us straight down. But as well as this force, we also experience force due to the motion of the rollercoaster, in particular the force due to its acceleration. This acceleration can be positive when the ride is speeding up, or negative as the ride slows down. The really wicked rollercoasters can also throw a few forces at you in a sideways direction.

The resulting sensation is one of being pushed in a variety of directions, depending on the direction of the acceleration. But we don't feel both forces separately: we feel their combined effect, and a rollercoaster takes advantage of this. So, if we are accelerating uphill fast we feel both forces (gravity plus the effect of the ride) in the same direction, and feel extra heavy. If we are falling quickly, then the effect of the forces can cancel each other out, so we feel weightless.

This apparent change in weight provides the strange sensations we feel. When we are hurtling downhill, all the loosely connected parts inside our body are being accelerated individually, which feels funny. In a fast fall there is virtually no downward force when the two forces are combined, so our stomach may feel weightless, giving us that sinking feeling.

Why can't **people tickle** themselves?

Tickling excites the fine nerve endings beneath the surface of the skin. It makes some people laugh, while others recoil from the touch.

How ticklish a tickle feels can depend on who's doing the tickling! Recent studies have shown a difference between brain scans of people being tickled by another person, compared to when they tickle themselves. In the case of self-tickling, it seems that your brain sees it coming and tells you to ignore it. Brain scans of self-ticklers show that the cerebellum, a part of the brain used in planning, sends urgent messages to another part of the brain warning it that a sensation is on the way.

However, we have to have control over this sensation; for example, life would be impossible if the soles of our feet tickled every time we put them to the ground. So the brain deals with this and sorts out the important stimuli from the ones which don't matter very much.

Darwin the evolutionist was interested in the phenomenon of tickles. He realized that a victim of tickling would squirm in order to get vulnerable parts of the body away from the stimulus. He thought this was an evolutionary mechanism to protect us against predators. Interestingly, the pleasure from being tickled increases the older you become.

Why do you **sometimes** get that 'falling off the edge of a cliff' feeling **while** you're asleep?

When you are asleep, the muscles in your arms and legs are actually paralysed. If they weren't, you would act out your dreams! Imagine the results of that. The problem is that this 'paralysis' doesn't always happen at exactly the same time as you fall asleep.

9

Number
Crunching

Starting at Zero

Who discovered **zero?**

The Babylonians had a numerical system based on the number 60, but they had no zero and instead simply left a space. The problem they soon came across was how to write 10, 100, 101, etc., without a zero. So, they were forced to invent a zero to get around this problem, and used a symbol which looked rather like an angle sign that we use today. It appeared about 500 BC. Strictly speaking, though, this was more of a marker than an actual 'number' zero.

The use of zero as a number in its own right would have been introduced when negative numbers were being used, and there is evidence of their use in India in the seventh century. The mathematician and astronomer Bhramagupta was the first to come up with rules governing zero and negative numbers when he stated, 'The sum of zero and a negative number is negative, the sum of a positive number and zero is positive, the sum of zero and zero is zero.' Not all his understandings of the concept of zero were correct, but this represented a major step forward in mathematical understanding.

Is **zero** considered an **even number?**

A good question! The definition of an even number is one that when divided by 2 leaves no remainder. This would seem to suggest that since zero divided by 2 is zero, with no remainder, then zero is in fact an even number.

However, one maths dictionary that I looked in said that even numbers were part of the infinite series of integers 2, 4, 6, 8, etc. It consciously did not include zero in the series. But I think that the dictionary was being overly restricted in its definition since 0 is an integer and most mathematicians would agree that zero is an even number.

I've heard that if you have to pick a **prize** (such as by **opening** a choice of doors only one of **which** has the **lucky** prize behind it) you stand a better chance of **winning** if you **change** your mind after making the first **choice.** Is that true?

Yes! Look at it this way: you're on a quiz show where there are three doors marked A, B and C. Behind two of the doors are goats, behind the other is a new car. If you choose door A, and then the quiz host shows you that door B has a goat behind it and asks if you want to change your mind, you have a greater chance of winning the car if you say yes and choose door C. Weird but true.

It's because the first time you choose you have a 2-in-3 chance of picking the goat, compared to a 1-in-3 chance of picking the car. In other words you have a greater chance of picking the wrong door. So 6 times out of 9 the door you pick will have a goat behind it. If you are then told which of the other doors has the other goat behind it, for those 6 times you can safely choose the final, third door knowing full well that the new car is behind it. Of course, 3 times out of 9 you'll actually have picked the car in the first place, so changing your mind will actually guarantee that you pick a goat.

Unfortunately the problem with chance and probability is that you never know which of the possibilities you're going to pick. Try it out with three egg cups and one pound coin. Hide the pound coin, ask a volunteer from the audience to choose an egg cup. Show them one without the pound underneath, and ask if they want to change their mind. Keep a tally recording if they won the coin or not. You should soon see that if they change their minds, they win more times. Of course if you want to win a goat, you should stick with your first choice.

Would I be better off
doing the **National Lottery**
in which you only have to pick
6 **numbers** out of 49?
How do I calculate the chances
of getting that **right?**

There are 49 numbers to choose from and you have to get 6 right to win the big prize. The chances of getting the first ball

right are 6 in 49. This is the chance that a ball drawn from 49 matches one of your previously chosen 6.

The chances of getting the second ball right are 5 in 48. The chances of getting the third ball right are 4 in 47. And so on down to 1 in 44.

When you have a probability of something happening AND another probability of something else happening, you multiply the numbers together to get the total probability of both events. So, to get all the numbers right you have to multiply 6/49 x 5/48 x 4/47 x 3/46 x 2/45 x 1/44. This gives a chance of 1/13983816.

So, there is 1 chance in 13,983,816 of winning the jackpot in the lottery. I'd stick with quiz shows.

Do you have less chance of winning the **lottery** **twice** in a row or twice in **your** lifetime?

It would make it easier if you simplify the question and ask about the probability of winning if you play every week for eleven weeks.

Let's say that the probability of winning is 1 in 10, and you win the first week. The probability of you winning again is 1 in 10 each day and so you are likely to win on one of the following days. If you only play the next week, your probability of winning would still be 1 in 10. So your chances of winning the next week are the same as your chances of winning on the 7th day or the 11th day.

Clearly the more you play the more opportunity you have for the probability to fall in your favour but it does not change the probability.

What is a **prime** number, and how can I **tell** if a **number** is prime or not?

A prime number is an integer (a whole number) larger than one that is divisible only by one and by itself. If it's not a prime number, it's called a non-prime number or composite number. So, 3, 5, 7 and 11 are primes – and the rest you can work out for yourself, perhaps. Be warned, though, finding out if a number is prime or not isn't all that easy once the numbers start to get past the hundreds. The mathematician Fermat (of the famous 'Last Theorem') suggested that if p is a prime number, then a to the power of (p-1), minus one, will be divisible by p. If it is not divisible by p, then it is not a prime. But if it is, it's still not necessarily a prime number, as some composite numbers also give a result that is divisible by p. Such numbers are known as pseudo-primes.

So how **many** **prime** numbers are there in **total?**

There is no limit to prime numbers. They are truly infinite simply because they are whole numbers and there is no limit to those either. But what if you argue that prime numbers are a subset of whole numbers, so the prime numbers must run out before the whole numbers do? It's an interesting argument, but not valid. Because of the definition of infinity there will never, ever, be an end to the list of whole numbers: they will go on and on for ever, and so will the primes.

What is 'pi' and what's so special about it?

There can't be many people who haven't been taught that the ratio of the circumference of any circle to its diameter, no matter what size the circle, will always be the same figure – pi. Usually its value is taken as 3.14, but pi is what's known as an irrational number which never ends. For example, a quarter is precisely 0.25, but a sixth doesn't end after a few numbers and becomes 0.166666 ... and on it goes, for ever. So does pi. If you want to take it to 18 decimal places then it comes out at 3.141592653589793238, but it certainly doesn't end there.

The importance of pi has been understood for over four thousand years: both the Babylonians and the Egyptians worked out that the circumference of a circle and its diameter were always related by this constant number called pi, although the value they put on it was approximate by modern standards. In very ancient times 3 was used as the approximate value of pi, and not until Archimedes came along in the third century BC does there seem to have been a scientific effort to compute it; he reached a figure equivalent to about 3.14. By the early sixth century AD Chinese and Indian mathematicians had independently confirmed or improved the number of decimal places. Early in the twentieth century the Indian mathematical genius Srinivasa Ramanujan developed ways of calculating pi that were so efficient that they have been incorporated into computer algorithms, permitting expressions of pi in millions of digits. So far, pi has been calculated by computers to 200 *billion* decimal places, and it's still going strong.

Who **invented** the equals sign?

This is one of the great unanswered questions, but we do have some good clues as to its origin. There's a scroll in the British Museum called the Rhind Papyrus which is about half a metre wide and nearly five-and-a-half metres long. Although several important fragments are missing, it forms the basis of our knowledge of Egyptian maths and includes the earliest known symbols of mathematical operations, including the earliest known sign that somewhat resembles our modern equals sign. However, it is placed at the end of the problem at hand, and although it is shaped a little like our equals sign the ends of the two parallel lines are joined, and in the centre of the top line is a small w touching the line. So that may or may not be an ancestor of our present equals sign.

The equals sign we use today would seem to have been introduced widely by Robert Recorde in his textbook of algebra, *The Whetstone of Witte*, written in 1557. He said, 'I will sette as I doe often in woorke use, a paire of parralles, or Gemowe lines of one lengthe, thus: ==, bicause noe 2 thynges can be moare equalle.'

Is there a special name for 1,000 trillion?

In the traditional British numbering system, the number 1,000 trillion is referred to as '1,000 trillion': there is no special name for it. A thousand million, on the other hand, was sometimes called a 'milliard' before the American use of 'billion' became usual.

However, in the American numbering system, which is increasingly used internationally, a thousand trillion is a quadrillion. But this is smaller than the British 1,000 trillion, because the US trillion is smaller: the old British trillion = a million billion = a million million million, so a thousand trillion = 1,000,000,000,000,000,000,000 (21 noughts) which is a sextillion in US numbering. The US trillion = a thousand billion = a thousand thousand million, so a thousand trillion = a quadrillion = 1,000,000,000,000,000 (15 noughts).

10

Can You Just Explain . . .?

Why do **people** say 'as easy as **falling** off a log'? **How** easy **is it** to fall off a **log?**

I am assuming this to be a serious question, to which you will get a full and extremely serious answer. Several factors come into play, all of which will interact in some way to facilitate or hinder your falling off the log you have in mind.

To be considered are your sense of balance, the log's slipperiness and how well your shoes grip the log. The size of the log also comes into play – very small logs will be less stable than larger logs. Small ones tend to rotate more easily encouraging falling because they require greater balancing skills.

To test your sense of balance try standing on one leg. If you really want to fall off easily, ruin your sense of balance by spinning round many times before you attempt standing on the log.

Different log properties will affect the slipperiness of the log. For example, the bark of a silver birch is much smoother than that of an oak tree and so will reduce grip.

The amount of algal growth on the log should also be a factor and this is determined by the moisture content of the wood – the wetter the log the more likely you are to slip off it.

Finally, the shoes you are wearing are important. If they have no grip and little surface area in contact with the log you will fall more easily. Any flat shoe with good tread will allow the water to drain through the tread, preventing a film of water forming between the shoe and the surface of the log, increasing the friction between you and the log. To make falling off the log even easier, try wearing stilettos – no tread, tiny surface area and serious hindrance of balance.

Then there's your centre of gravity, which is usually somewhere near your belly button. If you lean over so that a line drawn vertically from your centre of gravity to the ground falls outside your base (which means it does not pass through your feet) then you will fall over, unless, of course, you move your feet. The taller you are, the higher your centre of gravity and the less tipping that needs to be done before you become unstable. Obviously larger feet increase your base size and make you more stable. So a tall person with small feet will fall off a log more easily than a short person with big feet.

Imagine you're looking
into the rear-view mirror
while stationary
at traffic lights.
You **see a** car speeding up
behind you. He hasn't seen the
lights. He's going to **hit you!**
Is it **better** to leave your
brakes on,
or quickly take **them off** to minimize
the damage?
Or doesn't it **matter?**

Personally, I'm voting for the release-the-brakes option.

The impact is what causes the damage because when an object hits another object the momentum of the first object has to be reduced to zero. This can happen over a long or a short period of time.

The change in momentum is the product of the force and the time during which the object experiences that force. So, for a given momentum the longer the time of impact the smaller the force felt, and the shorter the time of impact the larger the force. It's like jumping up and down: if you bend your knees you increase the length of time you impact the floor and it is painless. If you keep your knees rigid the jump stops abruptly and it hurts. Take the brakes off for less of an impact.

Why are there **dimples** on golf **balls?**

They create turbulence round the ball, without which you could get a vortex developing behind the ball which would act as a pulling force, dragging it backwards and slowing it down. Dimpled balls go faster.

There's also Bernoulli's Principle to consider. Imagine a ball travelling from right to left in front of you. The dimples trap air, and if the ball is spinning in a clockwise manner the dimples on the top surface will speed up the air (since they are rotating with the airflow) while those on the bottom will be going against it, decreasing the speed. Bernoulli's Principle says that when air speeds up it reduces in pressure. So, you get a reduced pressure on the top surface of the ball, and an increased pressure on the bottom surface. That's called lift! If you can get the ball to spin in the opposite direction, as tennis players can, you will increase the pressure on the top and decrease the pressure on the bottom and make the ball dip more sharply.

Why **doesn't** this work?
If you take off in a **helicopter**
and just **hover** in one position
for long enough,
letting the Earth **rotate** underneath
you, **wouldn't** you have travelled
quite a **distance**
without
going anywhere?

You *seriously* think that if you hover over London long enough, by the time you come down you'll be in Paris? True, the Earth is spinning, and Newton is also turning in his grave at your lack of appreciation of his principle of the conservation of energy.

If you took off vertically in a helicopter, even with enough fuel, you would not orbit the Earth. It's all to do with momentum. While you are standing on the surface of the Earth, which is moving at 530km/hour at the equator, you have angular momentum by virtue of being in contact with the Earth. Momentum is always conserved, according to Newton. In other words, *momentum is neither created nor destroyed, but only changed through the action of forces as described by Newton's laws of motion.* Simply by jumping in the air you cannot destroy your body's momentum, and if you want to prove it then all you have to do is jump on the spot. Does the Earth move under you? Do you land on the same place? Of course you do.

The same happens in a helicopter, which also has angular momentum that cannot be changed other than by the action of forces. If you want to move the helicopter, you've got to open the throttles to provide those forces and then you'll travel. But by hovering you'll get nowhere at all.

If **a bee** found its way
on to a bus
and **hovered**
somewhere **near** the front,
and then the **bus**
drove away,
would the bee **stay** at the front
or be **swept**
to the back?

What applies to you also applies to the bee. As the bus speeds away, all the air in the bus isn't swept to the back, is it? This is because as the bus accelerates, the air inside it accelerates at the same rate, and so the relative motion between the two is unchanged. The bee is hovering relative to the air inside the bus and will continue to do so. Since the air inside the bus isn't moving relative to the bus, then the bee stays in the same place.

So I'm **beginning** to think
that when things hover,
nothing much
seems to happen –
helicopters don't move,
bees stay where they are.
So, if you've got a **budgie**
in a cage, would the cage weigh less
if the **budgie**
took off from its perch
and flew **around?**

Amazingly, yes, the total weight *would* go down. When the budgie is sitting on its perch, the total weight is a combination of the weight of the cage and the weight of the bird. When the bird launches itself into flight, its weight no longer has any effect on the scales the cage is on and so the cage would weigh less.

However, if the bird were to be in an airtight container (severe warning: *don't* try this at home) the weight would remain the same. This is because as the budgie flies, it beats its wings and every stroke of its wings pushes a column of air downwards with a force equal to its weight. In a cage with open sides, this air escapes so has no effect on the weight. We're back to Newton's laws and the conservation of energy again.

Could I achieve the same **effect** –
I mean
appear to **weigh** less
if I **stand** with just one foot
on the scales, **instead** of two?

Just placing one foot on the scales is no answer to the eternal weight loss problem.

Weight is a measure of force, and force can be represented as pressure x area. So, if you stand on the scales with one foot instead of two, what's happening is that although the area that is in contact with the scales has been reduced, the pressure acting on that area has increased proportionately, so the force – your weight – remains the same.

What if I had **two sets** of scales,
and put **one** foot on each:
would it be **possible**
to alter **my weight** by rocking
from side to **side?**

If you had two scales with one foot on each, the weight that is measured on each scale will be dependent on the area of contact between the foot and the pressure exerted on the scale, but the combined weights should add up to your normal weight. If you try this and the weights don't quite add up, it will be due to losses in the scales and not to any flaw in the principle. Try it.

11

Brainstorms

From Grey Matter to
Bread Crusts

I've been **told**
that people **only** use a
small part of their
brain, about
10 per cent. Is that **true?**

Let's just remember what a remarkable piece of kit we have beneath our skulls. If you've read from the beginning of this book you have given your brain a fair amount of exercise. You have taken in a huge amount of information, given much thought to the questions and answers, and all this has happened inside your head within an organ that weighs a mere three pounds. Mind you, it's a pretty packed three pounds and a hungry one.

There are 100 billion neurons in the average human brain, which is 20 times the population of the earth. These cells, the neurons, are both broadcasters and receivers of information which take in minute electrical currents and forwarding them as required. They are the basis of the entire nervous system. If you prick you finger, for example, it will be a flow of information through your neurons from your finger tip to your brain and back again that will eventually tell you it hurts.

Of all the organs in the body, the brain is a major consumer of energy. All those neurons are like small batteries, and when they pass on their electric charge it has to be replaced. All that energy has to come from somewhere and it has been calculated that 20-30% of the body's total energy is consumed by the brain alone. Look at it another way – over a fifth of everything on your plate is going to keeping your brain working.

The idea that the vast majority of the brain is unused, is a myth. Think about it: the human body is a finely evolved piece of machinery which has reduced itself to the most efficient state of operation. Why would it be running something as greedy as an idle brain for no good reason? This is what the evolutionists would argue.

But it is true that not all the brain is working all of the time. Bits of the brain will drop into an idle mode if you're not using them, as when you're asleep or listening to music with your eyes closed. It's possible that while that's happening you're only using a small portion of your brain power. But if the phone rings and you jump up to answer it, all those neurons will burst into life and not many of them will remain fast asleep.

Is the brain **cleverer** at some things rather than others? For example, is it better at **helping** us **recognise** taste because we only need **one mouth** for our food? But when it comes to **hearing** and **seeing** we've got two ears and two eyes. Is the brain **not** quite so good at these jobs and that's **why** it needs **two inputs?**

It's the other way round. It's because our brain is so clever that we have evolved to have two ears and two eyes. It's all to do with knowing where we are in relation to other people and objects, which was (and still is) a vital survival skill. It's no good being able to see the edge of a cliff, for example, if you can't tell how far away it is.

Our ears and eyes are similar, in a sense, because each of them provides us with a slightly different image. If you hold something close up and close one eye, and then the other, you see a slightly different picture in each eye. The same is true of sounds, and it is this slight difference in sight and hearing which helps us to decide where any particular object is placed or sound is coming from. Our mouths, on the other hand, play no part in direction finding which may be why we only have one of them. Anyway, imagine life with two mouths. Deciding which one to put the food in might be interesting – one for hot, the other for cold? It would also mean that our voices could appear from two places at once which would make the direction finding job of the ears even harder.

Because **Albert Einstein** was so clever, did that **mean** he had a **bigger brain** than most **people?**

Brain power has nothing to do with size, and Einstein proves it. Einstein was a smallish man, certainly no giant, and so had a smaller brain to match – in fact, it was smaller than average. What made Einstein particularly clever, and what makes most people cleverer than others, is the ability of the neurons in their brains to make connections. The more connections you can make, the cleverer you are. The physical size of your brain makes no difference. Remember, you have 20 times more neurons in your head than there are people in the world and the more you get the neurons in that heavily populated head talking to each other, the cleverer you will be. Neurons, incidentally, are some of the oldest cells in our body, some of them lasting a lifetime. To give you an idea of the number of neurons in your brain, if you were to place one sheet of paper on a pile for each of your neurons, you would end up with a tower of paper 5,500 miles high.

People talk about having 'brainwaves' when they've had a good idea. What are brainwaves?

Your brain is experiencing brainwaves all the time, but that doesn't mean it's always overflowing with good ideas. Electrical waves travel between the neurons continuously. Some of these waves are at a higher frequency than others and it is the frequency which determines the job the brain waves perform. There are generally reckoned to be four brainwaves: alpha, beta, theta and delta.

If you're busy, thinking hard, trying to work something out or answering questions under pressure, you are generating beta waves of 20–40 cycles per second. Suppose you've just taken part in a quiz and your head was flooding with beta waves, as soon as you started to relax the brainwaves would slow to 10–14 cycles per second and these would become alpha waves. If you've just done something you found quite difficult, sit down and take a deep breath to think it through, you're now making alpha waves in your head. They are the product of a relaxed mind.

If you then slip into an even lower mental gear, your brain waves drop to about 5 cycles per second and as you start to daydream the theta waves take over. If people say to you, 'you look a million miles away,' that's because your head is now filled with theta waves. There is something very relaxing about being in a theta state which is why some people always claim to have their best ideas in the bath.

Once the brainwaves drop to 2 or 3 cycles per second, you've found the delta state. The chances are that by now you are in a deep, dreamless and refreshing sleep and the lower the frequency the deeper the sleep. But don't let those brainwaves drop to zero – that's reserved for the brain dead.

When I get a **headache,** it **sometimes** feels as **though** my brain is **bursting** out of my **head**. Is that why my brain feels so **painful?**

Actually, your brain cannot feel any pain at all. I don't recommend you ask a neurosurgeon to take his scalpel to your brain, but if he were to cut away at it you would not feel a thing even if you had no anaesthetic. In fact, brain operations are sometimes performed with the patient conscious, and by asking questions – such as, can you hear and see properly? – the surgeon can determine the precise area of the brain he is working in. Nature has ensured that our brain is well protected by our skulls so it doesn't feel pressure and pain like some other parts of the body. Because you've got a headache, it doesn't mean your brain is hurting.

Instead, think of a headache as a warning flag which is telling you that something, somewhere is not as it should be in your body. It can be hunger, a hangover, tiredness or muscular. All these can act as a trigger which sends alert messages to your brain.

The effect of these signals is often to cause expansion of the blood vessels around the brain and as these vessels swell they also send pain messages to the brain which gives you the symptoms of a headache as well as the pressured feeling inside your skull.

I sometimes get a **headache** if I eat too **large** a lump of **cold ice cream**. What's **happening** there?

You'd think it would be quite simple to explain this common experience, but science has not yet come up with a final answer. They only thing they're sure about is that almost everyone suffers from it at one time or another, and it doesn't mean there's anything wrong with you. It is assumed that the cold stimulation of nerves within your sinuses, which are right under the bone in your forehead, provides a short cut to the nerve pathways which have a direct line to the pain centres in the brain, from which spread the headache signals. These sinuses are exceptionally sensitive to cold. It can really hurt while it lasts, but it's never for long. Eat smaller lumps next time!

I'd heard that we **can't grow** new brains **cells**, and when they **die** they're **lost** forever. Is **that** why we are sometimes not so **bright** as we get **older?**

It's true that your entire nervous system is formed while you are still in your mother's womb, and it's the nervous system you'll have for the rest of your life. In the nine months before your birth you are growing neurons at an unbelievable 2.5 million every minute, and these have got to last you because you won't grow any more. Of course, as you grow up your brain will get bigger but that's not because you are producing more neurons – it's because the cells themselves are getting bigger. The more you exercise your muscle cells the larger they get, and the same is true of your brain cells.

But there's nothing you can do to stop neurons disappearing. As soon as your teenage years are over, they start to die and they vanish at the astonishing rate of 50,000 every day. By the time you're in your eighties, you've lost 10 per cent of them. You can make up for that, though, because neurons respond to training and form extra links with other neurons if encouraged to do so. Think of your brain as a muscle, and keep it fit.

People sometimes call you a **bighead** if you've **done** something that **shows** you to be **cleverer** than they are. Is a big head the sign of an **intelligent** person?

Not at all. It doesn't matter in the slightest how big your brain is as Einstein proved (see above). Your brain could be as big a bath tub but that wouldn't mean you were cleverer than someone with a brain the size of a kitchen sink. What makes you clever is the number of connections the brain cells make with each other, so size really doesn't matter.

I've **heard** that what we think of as **our** brain is actually **two brains** sitting side by side and doing **different** jobs. Is that **true**?

If you want to think of it that way, then we do have two brains called the right side and left side, although scientists would say that they're really one brain with a large fold down the middle and each side is connected to the other by a thick cable of nerves. The two sides of the brain are responsible for different jobs. The left side does the reading, writing and arithmetic, as well as controlling speech. Anything that has to do with a logical way of behaving or thinking is the job of the left side. The right side, however, controls how you see things and appreciate them. Intuition also lives here, which is one of the reasons that the right side of the brain is seen as the creative half.

If you have a brain dominated by the left side, you tend to become a scientist, a judge, banker or librarian because you like thinking in an orderly way. Right-brained people tend to be good at sports and arts, enjoy fantasy, write novels and love cats (for some reason).

There is some research which says that most males have a more powerful right side, and most females a dominant left side (see grey matter, below). You can go and argue that amongst yourselves.

How **fast** does the **human** brain operate? Can you measure it in **mega-hertz**, like you can a **computer**?

It is very difficult to compare a human brain to a computer because they operate in completely different ways. A computer operates in a linear way, using its processors to perform one function at a time. Part of its speed comes from its ability to use its vast memory. But the brain contains vast numbers – billions – of neurons which are the human form of processors, and these are interlinked and work together to perform brain functions, unlike a computer which is doing it in a methodical, sequential way, albeit quite fast. This is why the brain is much faster at recognising objects, colours and sounds for example, whereas a basic calculator can perform addition and multiplication at far higher speeds than the human brain. The brain and the computer are completely different tools.

But our **brains** have some **memory**. Is it **possible** to say how much? My computer has a **50 Gigabyte** memory. **How** much does **my** brain have?

It's one of those questions to which there is no answer because you are not comparing like with like. Information in a computer is held in discrete locations, but the brain works by creating networks between its neurons as required to perform its various functions. So, with a computer it is easy to measure memory because you can count the discrete locations and measure them in bytes. But the brain appears to be constantly changing its internal workings as varying demands are made upon it. In other words, there's no way of counting it and so it's impossible to give a figure.

Someone once told me that we have the largest brains of any creature on the planet. That can't be true, can it?

Not strictly true, but I can guess how the confusion arose. Without doubt the massive blue whale, the largest creature on the planet, has the heaviest brain. It weighs about 6 kg to control a body which could weigh as much as 25 elephants. We are puny by comparison, but the human brain is larger in relation to our weight than that of any other living thing on earth, which is why the blue whales don't rule the world.

How long will our brain survive if they don't get a proper supply of oxygen?

Not very long at all. The precise time will, of course, depend on a number of factors but you can reckon significant damage would be done to your neurons after ten minutes. That is why doctors are always so keen to make sure emergency patients are breathing and getting a proper supply of oxygen before they consider any other treatments which might be required.

In fact, the brain is an oxygen-hungry beast and uses 20 per cent of all the oxygen we breathe in.

Why do **people** talk about the **brain** as being 'grey matter'?

Parts of the brain are grey, about 40%, the remaining 60% is white, although women differ from men. Women have smaller brains, generally, but have more grey matter in them than men. It is the grey matter which performs the brain's processing functions, and the white matter acts as the conductor which moves the information around. Although it's a bit dangerous to draw close links between brains and computers, you could say that the grey matter did the job of the microchips, and the white matter was the printed circuit that joined them together.

It is thought that the different ratios of grey and white matter in male and female brains accounts for the fact that women are often better communicators, while men may have enhanced spatial awareness. This could be why it needs a man to read a map, and a woman to explain why they got lost in the first place. Recent research has suggested that the amount of white matter in our brains may also affect our ability to tell lies and deceive. As it takes a lot more careful thinking to tell a successful fib, scientists now believe that the white matter in the brain provides the tool which helps liars think more quickly, because the more white matter there is the more rapid the connections between the processors in the grey matter can be made. The instinct to behave ourselves and not tell lies is also found in the grey matter, and habitual liars tend to have less grey matter, making their problem even worse.

The grey matter is hungry stuff - 94% of the brains oxygen is used to feed it.

My **mother** always used to say that **eating** bread **crusts** makes your hair **curl**, and eating **fish** made you **brainy**. True?

There could well be foods which help your brain to work more efficiently, and fish is certainly one of them, especially oily fish such as tuna, herring and mackerel. But the brain's main requirement is for a regular supply of glucose and if you skip breakfast, for example, you are never going to kick-start your brain to face the day. Research has shown that having no breakfast makes school work harder throughout the morning. But beware what kind of glucose you take. Fizzy drinks and chocolate bars do not work as well as 'proper food', such as eggs, bread or cereals. Eggs are particularly good because they promote the production of chemicals which help information transfer between neurons. Junk food not only helps you pile on the pounds, but in a controlled experiment rats fed on junk food found it much more difficult to find their way round a maze than when fed a normal diet. So there are foods which you could say make you brainy.

Science has yet to consider the hair curling properties of bread crusts.

Index